The Elasticity of Life

Series Editor
Jean-Charles Pomerol

The Elasticity of Life

From Tissues to Humans

Pascal Sommer
Romain Debret

WILEY

First published 2023 in Great Britain and the United States by ISTE Ltd and John Wiley & Sons, Inc.

Apart from any fair dealing for the purposes of research or private study, or criticism or review, as permitted under the Copyright, Designs and Patents Act 1988, this publication may only be reproduced, stored or transmitted, in any form or by any means, with the prior permission in writing of the publishers, or in the case of reprographic reproduction in accordance with the terms and licenses issued by the CLA. Enquiries concerning reproduction outside these terms should be sent to the publishers at the undermentioned address:

ISTE Ltd
27-37 St George's Road
London SW19 4EU
UK

www.iste.co.uk

John Wiley & Sons, Inc.
111 River Street
Hoboken, NJ 07030
USA

www.wiley.com

© ISTE Ltd 2023

The rights of Pascal Sommer and Romain Debret to be identified as the authors of this work have been asserted by them in accordance with the Copyright, Designs and Patents Act 1988.

Any opinions, findings, and conclusions or recommendations expressed in this material are those of the author(s), contributor(s) or editor(s) and do not necessarily reflect the views of ISTE Group.

Library of Congress Control Number: 2023935278

British Library Cataloguing-in-Publication Data
A CIP record for this book is available from the British Library
ISBN 978-1-78630-907-5

Contents

Foreword ix
Yves RÉMOND

Preface xiii

Acknowledgments xv

Introduction xix

Part 1. Fantastic Elastic Capital 1

Chapter 1. Strengths and Weaknesses of the Elastic Human 3

    1.1. Introduction. 3
    1.2. Longevity and elasticity 4
    1.3. Disasters. 4
        1.3.1. Fibrosis of infectious origin. 5
        1.3.2. The great pandemics 6
    1.4. Cutaneous elastic capital 8
        1.4.1. Introduction 9
        1.4.2. Wrinkles and scars 9
        1.4.3. Beautiful skin 11
        1.4.4. The elastic network of the dermis 13
    1.5. Loose skin and cutis laxa. 15

1.6. What is missing and malfunctioning in these fiber diseases?. 19
1.7. What structures and strengthens the elastic system 20
   1.7.1. The mechanical actors 21
   1.7.2. Cross-links. 23
   1.7.3. Calcification 25

Chapter 2. Elastic Capital, Air, Water and Other Fluids 29

2.1. Introduction. 30
2.2. Respiration 30
   2.2.1. The breathing cycle. 31
2.3. The 12/8 of perfect blood pressure!. 34
2.4. Cellular respiration and energy 37
   2.4.1. ATP, our universal source of energy 37
   2.4.2. Our mitochondrial battery. 39
2.5. The logistics of digestion. 40
   2.5.1. The intestinal walls: a history of pushing 40
   2.5.2. The lazy lymph 41
   2.5.3. The kidneys: a story of water and blood 42
2.6. Vascular dilation and constriction 43
   2.6.1. The SARS-CoV-2 gateway and blood pressure 44
2.7. Sugar logistics 45
   2.7.1. The liver and blood sugar 45
   2.7.2. The kidneys and liver under sugar pressure 47
2.8. The perineal set and terminal delivery logistics 47
   2.8.1. Cross-linking and vaginal prolapse 48
   2.8.2. Mechanical adjustment to pregnancy 49
2.9. The microbiota and its body bioreactor 50
2.10. Conclusion 52

Chapter 3. Elasticity and the Senses 55

3.1. Introduction. 55
3.2. Singing and dancing. 56
3.3. Light transmission and elasticity 59
3.4. Auditory transmission and elasticity 60
3.5. Olfaction 61
3.6. Taste. 62
3.7. Touch and proprioception 62
3.8. Elasticity and the peripheral nervous system 64

Part 2. The Four Challenges of the Elastic Human 67

Chapter 4. The First Challenge for the Elastic Human: Mechanical Stress Management 69

    4.1. Introduction. 70
    4.2. Stress of elastic structures 70
    4.3. Stresses on organs and fluids 73
    4.4. Genetic stress. 75
    4.5. Stress and epigenetics. 78
    4.6. Pharmacology and stress 79

Chapter 5. The Second Challenge for the Elastic Human: The Management of Food and Inputs 83

    5.1. Introduction. 83
    5.2. Elastic capital and phytotherapy 84
    5.3. Elastic capital and dill. 85
    5.4. Epigenetics and marjoram 87
    5.5. Adopting a plant. 89
    5.6. Elastic capital and protein restriction. 90
    5.7. Elastic capital and a reasoned diet 91
    5.8. Glycine on the menu 93
    5.9. Elastic capital and an unreasonable diet 95
    5.10. Elastic capital and pollution 98

Chapter 6. The Third Challenge for the Elastic Human: Successful Life through Movement 103

    6.1. Introduction. 104
    6.2. The alchemy of movement. 105
    6.3. Elasticity at both ends of life. 106
    6.4. The reference frames of motion. 108
    6.5. Flattering the view and looking good. 110
    6.6. Feeling your proprioception to improve your posture 113
    6.7. Touching and stroking to promote elasticity 115
    6.8. Hearing and feeling in order to enjoy the song and movement. 117
    6.9. Conclusion: let us adopt a loop of "pro-elastic" postures 119

Chapter 7. The Fourth Challenge for the Elastic Human: Adopting the Spirit of the Laws of Elasticity ... 123

    7.1. Introduction... 124
    7.2. Cross-linking and knowledge ... 126
        7.2.1. Knowledge and the expert patient ... 127
        7.2.2. Knowledge and cyberchondria ... 129
    7.3. Resilience and the mechanics of the world ... 131
        7.3.1. Resilience and the elastic human... 131
        7.3.2. Resilience and society ... 132
    7.4. Compliance, ethics, law and elastic capital ... 135
        7.4.1. Compliance ... 135
        7.4.2. The elastic system and the technical system ... 136
        7.4.3. Mutual aid and the elastic system ... 137
        7.4.4. Culture and the elastic human ... 139
        7.4.5. Morality and the elastic human... 141
    7.5. Chapter summary ... 144

Conclusion... 147

Appendix... 155

References... 193

Index... 211

Foreword

After years of working together, Pascal Sommer and I organized a scientific conference in French at the headquarters of the Centre National de la Recherche Scientifique (CNRS) in Paris entitled "Repairing the Elastic Human: From the Fundamental Sciences to the Surgery Room". Pascal Sommer as a biologist researcher, a recognized specialist in tissue elasticity among others, and I as a mechanic of materials and biomaterials, and specialist in engineering sciences, had prepared this event within the framework of our responsibilities at the CNRS. The title of this conference was surprising for many and did not fail to attract attention. It was a success. Scientists of great talent explained the progress of knowledge on the subject and especially the stakes of this elastic human, the heart of Pascal Sommer's essay that I have the honor and the pleasure to comment on here. Questions were flying around: what could be the role of elasticity in the human body?

A few days later, we were guests of Fabienne Chauvière, journalist and producer of the program "Les Savanturiers" on France Inter, dedicated to scientific advances. The participation in radio programs, without being exceptional, was not something we did regularly. So, as soon as the program began, the journalist asked me point-blank if the title of our conference was not a bit provocative. Surprised by this incisive start, I gave a general and, it must be said, conventional answer, a little bewildered by the question. Pascal Sommer then took the floor and answered Fabienne Chauvière that without elasticity, she would not be alive, her heart would not beat, her lungs would not function and allow her to breathe, her blood would not flow for long in her arteries, and the movements of her body would be so impossible that her skin would crack and age would weaken the functionality in all parts of the body. The surprise was reversed, and the program got off to a good start. I think it was this answer, which lasted less than a minute, that showed me in a concentrated way how vital the elasticity of the human body was. The symposium had shown that the observation of this elasticity was now so well advanced and well documented that its role was taken very seriously, but its determinants were only imperfectly known. It was necessary to understand them at all human scales, from that of DNA or proteins to the macroscopic scale of tissues, passing through the scales of the various cellular or tissue components. But let us return to the mechanical description of elasticity, since I have the honor to introduce this concept in this book at the request of Pascal Sommer.

That elasticity is a property of inert matter has been known since the dawn of time. It was only conceptualized in the middle of the 17th century, because of a great English scientist named Robert Hooke. Then, the engineers and scientists who followed continued his study, and their results are the basis of all designs of objects, machines, buildings or vehicles that surround us. Its extension to the observation of the living is however much more recent. It is striking to see that this scientist who studied at Oxford, to whom we owe also the first acoustic telephone, had synthesized the elastic law of bodies by a sentence that has become famous: *Ut tensio sic vis* (elongation is proportional to the force). It is also interesting to note that Robert Hooke was also a pioneer in biology. We also owe him the word "cell" for living plant cells, which he was already observing under the microscope!

Today, elasticity is well understood and modeled in the mechanical domain. We distinguish different forms that all have in common the state of stress being linked to a material's state of strain at a point in space. We agree that these laws are very useful because they allow us to model, for example, the point of failure of an aircraft wing or the reactor vessel of a nuclear power plant. When mechanical behaviors are linear, with simple and unstructured materials, this elasticity is described by a law (called "Hooke's law") whose mathematical formulation is introduced in section A.1 in the Appendix. This organization is called isotropy when the mechanical properties of a segment are independent of the organization of the whole (this point is under discussion for characterizing the behavior of the universe).

And in the case of living tissue, you may ask? We have no precise information. It is obvious that biological media are rarely organized in a uniform way and that they all have an elasticity. The formulation of this elasticity, even if reduced, is very complex. In physical language, we qualify as anisotropic their organization, when organizational directions are imposed as in a venous wall or a bone. It is then necessary to have more mechanical constants to characterize and model them mathematically. Especially in warm-blooded animals whose elasticity may depend on temperature, which may require additional consideration for the laws of thermodynamics. It is likely, as with some elastic polymers, that we are in the presence of a mixture of several types of elasticity, but the question remains open and will probably depend on the scale at which we will place ourselves.

Finally, let us also recall that the material offers us interesting variants of elasticity:

First, viscoelasticity, which introduces delays in the elastic reaction of solicited matter. This is typically the case for living matter, where viscosities are at work at all scales. We can measure viscosity maps as well as stiffness maps of living tissues and derive useful clinical information from them. We are interested in this for muscular tissues with regard to Duchêne's disease, for example, or for the rigidity of tumors. The injection of hyaluronic acid in the joints responds to this need to locally increase viscoelasticity.

We could also talk about super-elasticity, which is typical of certain artificial materials used in the human body for their ability to have a shape memory. It is used for example to manufacture stents or surgical wire, which will be discussed in this book. This particular elasticity comes from a phase change of the material during solicitation. This phenomenon is then used to store particular strains and restore them, thus activating a memory effect. The most classical inert materials in this field are the Nickel–Titanium alloys found in some prostheses.

Finally, we mention the hyper-elasticity concept that has allowed us to understand and model the elastic behavior of many media and in particular biological media. The skin is an example. In this case, we use the data of a thermodynamic potential like the free energy from which a law of behavior is derived to characterize them.

This brief review of elasticity allows us to show that the elasticity of living tissues is part of a great history of solid and fluid mechanics, which started in the middle of the 17th century. It has expanded sufficiently to take into account many complex phenomena, including now the phenomena described by biology. It is in this context that Pascal Sommer's book is written, whose expertise in biological, human and social sciences perfectly illustrates the great diversity, but also the great complexity, that must be added to physics in order to deal with it.

Today, the developments of the elasticity of living organisms find a particularly fertile field of application in mechanobiology. The aim is to observe and then try to understand the effect of mechanics on biology. For example, we can thus direct the development of a stem cell or tumor cell by playing on the elasticity of the medium on which it is placed. This type of phenomenon, which is not yet fully understood, illustrates this surprising relationship. In another field, we can also see how the cells that are at the basis of bone remodeling and scarring are strongly influenced by gravity. An unfortunate consequence is the bone loss that many astronauts undergo during space flights in microgravity. This phenomenon, which has been studied extensively and is equivalent to osteoporosis, is not completely understood either.

We could multiply the examples over and over again; the field of action of the elasticity of living things is immense, and we are still far from having acquired the understanding and the knowledge of these numerous phenomena. Reading Pascal Sommer's book on this "elastic system", which has never been exposed at this level, will easily give readers the measure of the immensity of the biological phenomena associated with elastic behavior and their importance in our life. I am convinced that the next few years will be crucial for the knowledge in this field and that we will obtain new and useful results for humanity. Enjoy reading this book!

Yves RÉMOND
Université de Strasbourg

Preface

This book covers the elasticity of living things, from tissues to humans. It is a subject generally approached from a functional angle when the body becomes insurgent and manifests itself painfully. It is the register of physical limitations when mechanical functions become non-functional or missing. It is a register of morbidity when the lungs, the heart or the kidneys lose their necessary elasticity. It is also a parameter of capacities and feats when our elastic performances are confronted with our limits. The singularity of this elastic capital is being able to evaluate its own inexorable decline linked to age.

We often confine the problem of body elasticity to the appearance of wrinkles that mark our aging. However, everything that moves in the body is involved because elasticity is the means of absorbing the shocks of pressure or extension to quickly return to the normal situation. We need elasticity when we construct, run, sing, read, write, cook, eat, talk, watch, make love or give birth. At any moment, the elastic properties of our body condition our dynamics and our actions, whether it is in terms of bones, tendons, ligaments, fascias and muscles, but also in our lungs, hearts, intestines, kidneys or bladders. Elasticity is also necessary for sensoriality and is part of the register of behaviors; it is mobilized by the senses, sexuality and fertility. Finally, it is the register of the spirit and ethics when the repertoire of our elastic capital sets limits to our actions and societal proposals.

The subject of human elasticity is complex and interdisciplinary, but it is rarely presented in a comprehensive form. It is the intention of this book to position the concept of elasticity of living things at the heart of our lives, which is the reality. After a first part positioning the actors and the problems when elasticity is lacking, the book will address solutions to protect, maintain, reinforce or replace our elastic capital. As elasticity is omnipresent in society, its presentation is necessarily situated at the interfaces of experimental sciences (biology, chemistry, physics, mechanics) and human, social and societal sciences.

May 2023

Acknowledgments

The writing of this book was based on my interdisciplinary experience in the fields of biology and pharmacology of elastic tissues, aging, engineering for health and loss of autonomy. It was developed in response to the injunction of global pandemics, whether viral, dietary, behavioral or societal in origin.

Professor Yves Rémond, a specialist in the mechanics of materials and professor at the Université de Strasbourg, did me the honor of writing the foreword to this book and of contributing his expertise on the science of materials. His skills and enthusiasm have constantly enriched my much more limited knowledge of mechanics.

I thank Romain Debret, researcher at the CNRS, who participated in the work, especially in the appendices. His contribution on the development of drugs and devices is decisive to instruct an epigenetic regulation of altered elastic tissues. In addition, his involvement in patient association programs is exemplary. The 10 scientific figures were developed under his supervision by Anne-Lise Paris (www.in-graphidi.com) in the framework of a research action supported by the French National Research Agency (*Agence Nationale de la Recherche*, ANR-18-CE18-0001).

Because of her expertise in pharmacology and law, Valérie Siranyan's contribution was necessary to intelligibly formulate the ethical questions raised by the alteration of our elastic capital. A professor at the Université de Lyon 1, she advised me on the aspects relating more specifically to the human and social sciences. As an editor of books on national and international health regulations, her experience leads her to identify users' rights as a fundamental element of public health ethics. Her neologism of Handicracy, introduced during the course of this book, illustrates these notions well and was the subject of a joint symposium.

My work on elastic tissues has benefited from multiple resources provided by the CNRS, the universities of Lyon and Marseille and the Institut Pasteur in Lyon. I would like to thank all my past and present colleagues, as well as the French taxpayers and the donors of the Telethon who have contributed to the funding of our research. I would particularly like to thank my colleagues who passed through the Institut Pasteur in Lyon where this story really started, notably Michèle Chevallier, Alexis Desmoulière, Françoise Gérard, Claudine Gleyzal, Jean-Alexis Grimaud, Sylviane Guerret, Hugues Lortat-Jacob, Simone Peyrol, Mireille Raccurt and Sylvie Ricard-Blum.

The studies were carried out at the Institut de Biologie et Chimie des Protéines – CNRS – Université de Lyon, with notably Géraldine Aimond, Aurore Berthier, Romain Debret, Denise Eichenberger, Jean Farjanel, Bernard Font, Bérengère Fromy, David Hulmes, Caroline Reynaud, Dominique Sigaudo-Roussel, Jérôme Sohier and many others.

The work continues in Lyon under the aegis of Romain Debret and Jerome Sohier, as well as at the Institut des Sciences du Mouvement (CNRS – Aix Marseille Université) where I was particularly well received by Martine Pithioux, Patrick Chabrand, Jean-Louis Milan, Virginie Taillebot and Eric Berton, all experts in biomechanics.

Many doctoral students worked hard to give the best of themselves on this subject, and particularly Jérémy Boizot, Agnès Borel, Charbel Bouez, Thierry Brune, Valérie Cenizo, Stéphanie Claus, Leslie Laquièze, Claude Jourdan-Le Saux, Gabrielle Le Provost, Chloé Lorion, Carine Mainzer, Léa Moulin, Emmanuelle Noblesse, Marie Peraldi-Decitre, Floriane Pez, Sophie Sève and Laetitia Thomassin. They did a fantastic job!

Several collaborations have been important but certainly those with Odile Damour and her team at the skin substitute laboratory of the Edouard Herriot Hospital as well as those of Eric Perrier, Valérie André and Isabelle Orly from the company Coletica, now BASF Beauty Care Solutions France, which have offered the use of a unique model of human reconstructed skin.

The history of the *cutis laxa* owes its imprint to the Boiteux family, the originators of the international cutis laxa association. The anecdotes about the September 12, 2001, broadcast of the program "Ça se discute" organized by Jean-Luc Delarue are real. The constitution of a research consortium, some of whose funds came from the Telethon, was initiated by Christine Bodemer's team at the dermatology department of the Necker-Enfants Malades Hospital in Paris, then with Anne de Paepe's team at the Center for Medical Genetics Ghent. The complex notions of ethics of rare diseases were largely explained to me by Marie-Hélène Boucand.

My story on elasticity has been profoundly European, because of several programs financed in part by the European Commission that I contributed to setting up or leading with the participation of Daniela Quaglino at the Università degli Studi di Modena e Reggio Emilia in Modena, Julia Bujan at the University of Alcala de Henarès, Marie-Paule Jacob at the Bichat Hospital – INSERM – Paris, Brigida Bochicchio and Antonio Tamburro at the University of Potenza, Catherine Kielty at the University of Manchester, and Gilles Faury at the Hypoxia and Cardiovascular and Respiratory Pathophysiology Laboratory INSERM – Université de Grenoble.

I also thank Radovan Borojevic at the Federal University of Rio de Janeiro, Doris Germain at the Icahn School of Medicine at Mount Sinai – New York, José-Mauro Granjeiro at the Institute of Metrology of Brazil, Paulo Pinto Joazeiro, Instituto de Biologia, Unicamp – Campinas, Robert Mecham at Washington University – Saint Louis, Ursula Schlötzer-Schrehard at the Universitäts Augen klinikum in Erlangen, Philip Trackman at the Boston University School of Dental Medicine and Hassan Zahouani at the Ecole Centrale de Lyon, as well as Laurent Apert, Marielle Bouschbacher and Christelle Laurensou from the URGO laboratories, Carla Barichello and Noelle Remoué – Sohier from the Natura laboratory in Sao Paulo.

The rich contacts at the CNRS management and the CNRS mission for interdisciplinarity have made it possible to build a story where biological sciences, engineering sciences and human and social sciences converge. I would like to mention for their contribution Olga Allard, Philippe Bompard, Andréï Constantinescu, Luc Darrasse, Marie Gaille, Pierre Guillon, Christophe Jouffrais, Marie-Christine Lafarie-Frenot, Dominique Leguillon, Jean-Yves Marzin, Patrick Netter, Yves Rémond, Anne Renaud and Jean-Louis Vercher.

However, there were so many others like all the leaders of projects supported by the CNRS on the loss of sensoriality and autonomy who exploded the boundaries between sensoriality, movement and body mechanics as well as my colleagues from the Commission de Pharmacologie, Bio-ingénierie, Imagerie, Biotechnologies du Comité National de la Recherche Scientifique (CoNRS) that I had the honor to preside over. I illustrated my remarks on perception and disability with a few words on projects developed on sensory stimulation with Asaf Achraf, Malika Auvray, Pierre Ancet, Jérémy Danna, Leslie Decker, Coline Joufflineau, Agnès Robby-Brami, Fabrice Sarlegna and Jean-Luc Velay, on olfaction with Moustafa Bensafi, Liliane Borel, Jérôme Golebiowski and Norbert Noury, and on tactile perception with Marie-Ange Bueno, Marcel Crest and Betty Semail, on hearing and phonation with Lucie Bailly and Nathalie Henrich, on the control of assistive devices with Stéphane Buffat, Eric Campo, Loic Caroux, Jozina de Graaf, Nathanael Jarrassé, Laura Lemahieu, Noelle Lewis, Caroline Nicol and Nadine Vigouroux and finally on disability with Mai-Anh Ngo and Valérie Siranyan.

Some of my colleagues and friends who collaborated in this story have passed away and I would like to ask for them to be remembered: Stephen Baydanoff in Pleven, Jean-Louis Boiteux, Robert Frank, Léon Hirth, Jean-Paul Klein and Anne-Lise Pini in Strasbourg, Ladislas Robert in Paris, Olivier Toussaint in Namur, Yvonne Pasquale-Ronchetti in Modena and Tony Tamburro in Potenza.

The proofreading, corrections, foreword and discussion benefited from the meticulous work of Marie-Claude Boiteux and Mireille Tessier of the international cutis laxa association as well as Elyane Sommer. Any royalties from this book will be donated to associations involved in research on pathologies affecting elastic tissue.

And of course, I thank my family and friends, who had the patience to support me and read my work during the very long development of this book.

Pascal SOMMER

Introduction

Figure I.1. *The disasters of the elastic human. The intention of this first painting is to illustrate the weakness of a body abandoned by its elastic forces. It is inspired by Francisco Goya's engraving in his work entitled* The Disasters of War *illustrating the consequences of the invasion of Napoleonic troops in Spain. The painting* Gracias à la almorta *depicts the effects of the famine and the use of a bread substitute made of sweet pea flour. This plant contains a neurotoxin that can cause paralysis of the lower limbs and another toxin that prevents the proper formation of collagen and elastic fibers, with serious pathological consequences on the skeleton. The term lathyrism derives from the generic name of peas, including some species of* Lathyrus. *The engraving represents a family of which I have taken only three members with a woman unable to stand up to eat and drink*

For a color version of all the figures in this chapter, see www.iste.co.uk/sommer/elasticity.zip.

This book on the elasticity of living things aims to present a global vision of the elasticity of the human body. We could extend a large number of our reflections to all mammals. This vision of the elasticity of living things is omnipresent in art or culture because it is underlying the perfection of the body or its imperfections, innate or acquired with time. The representation of youth with beautiful smooth skin on a toned body, or of old age with craggy skin on weakened silhouettes constantly illustrates this throughout the history of painting. But the consequences of a lack of elasticity are even more strikingly illustrated in the opening figure, which depicts slumped figures, possibly in respiratory distress, unable to stand and begging for help to bring a simple jug of water to their mouths. I have freely used this engraving as a symbol for this book, just as I have used other famous painters to illustrate many parts of this book. Iconography abounds with these unfortunate people whose elasticity is degraded by diseases, infectious or not, capable of affecting many of our organs. So let us take some time and look at the elasticity of our human vehicle, which can be severely damaged by a toxin or a virus, but which is also damaged in a much more insidious way by other modern-day pandemics, linked to deleterious food and behaviors that concern us all.

Science has never been so much a part of our lives, nor has it provided information that can help us understand and manage our fantastic elasticity. The advances in science have undoubtedly reduced the toll of loss of autonomy associated with loss or defects in elasticity. The subject is topical because some of these defects have been seen in the consequences to the coronavirus pandemic since 2019, in the short or long term. This book carries the ambition of sharing knowledge on just about all the elastic capacities of the body that we unconsciously use when everything is going well and that become tyrannical when they are affected, due to simply being essential to life. There are many causes for the disruption of these capacities. The most obvious is related to aging, a universal cause. The loss of elasticity can also be amplified in times of threats and sanitary crises, such as during epidemics or viral, bacterial and parasitic pandemics, which humanity has seen, sees and will see again. We can die from a lack of elasticity, especially when a viral attack induces an exaggerated inflammatory response in the body and alters the mechanics of the respiratory and cardiovascular systems. Constraints are also very strongly imposed on our body mechanics due to the harmful effects of uncontrolled industrialization. They will probably be further amplified by future climate change.

The concept of elasticity of the body remains very vague, even if it imposes itself on our everyday realities, and to take it into account makes it possible to enrich playful or preventive behaviors or to support recommendations and prescriptions in multiple fields. It is our ambition to present it in a form that is accessible and systemic, in a functional and even performative approach. I know from experience, through the many questions I have been asked, that the notion of elasticity is only addressed occasionally and in critical situations. The first part describes the elasticity of the skin, tissues and organs as well as the body's structure (bones, tendons, ligaments and fascias) in a global manner. We will then move on to the aging and healing of the skin, as well as the calcification and strengthening of the structures that give the body its mechanical capacity. We will examine the difficult situation of children and adults who do not have adequate elastic capital. We will present the essential role of elasticity in the transport of gases, liquids and energy. We will see how this process is affected by respiratory and cardiovascular insufficiency, but also by excess food and drink (alcohol, sugar or meat). We will discuss how the five senses can depend on elasticity, to which we will add the sixth sense of proprioception. Finally, we will see how sexuality and fertility are intimately affected by elasticity.

This first description aims at linking the numerous aspects of human elasticity in a synthetic vision. Once these notions have been introduced, it will be a matter of considering the actions that promote or protect our elastic capital, knowing that these actions are often the result of expert considerations, affecting both our personal behavior and societal choices. It was therefore necessary to develop a vision that is based on both undeniable scientific foundations and strong social and societal representations. The range of this synthetic, interdisciplinary and timeless vision was therefore not obvious, and it is the almost mathematical analysis proposed by Jacques Ellul of an old text that offered me an adequate segmentation (Ellul and Rognon 2008). This is the analysis of the passage describing the four horsemen of the Book of Revelation, which we will consider only from the point of view of dialectical construction. The first horseman is emblematic of the spirit, whatever its substance; he rides a white horse, and he appears first but will really be active last. The second rider gallops on a fire-red stallion and spreads war and stress. The third rider rides a black steed and orchestrates famine. The fourth represents death and the immobility to come on his pale horse. Each of these four figures assumes a mission that can be both destructive and challenging. It is this delineation of four missions that attracted me and led me, by analogy, to orchestrate the presentation of four challenges for the elastic human. The first challenge consists of limiting the stress linked to the loss of elasticity as much as possible. In this book, we will describe the different manifestations of this stress that are often ignored, even by therapists. Advances in medicine and engineering are part of the fight against these stresses. This challenge is therefore a matter for specialists but also for expert patients; the more the former have a systemic view, the better their diagnosis will be; the more educated patients are, the better their dialogue with specialists will be. The second challenge is food, in the broad sense of the word, because it concerns all the inputs which are essential to the elasticity of living organisms, either by protecting and promoting our elastic capital, or by participating in the fight against pollution. It is certainly a matter for experts but also for common sense, in an economy that has lost some of its safeguards that often need to be reframed. The third challenge mobilizes our resources through movement associated with a thoughtful sensoriality, which seems obvious but is not useless to clarify. This need for action and perception is necessary to challenge the after-effects of a sedentary lifestyle or, conversely, the harmful effects of repetitive movements. The preservation of a good elastic capital, coupled with muscular and sensorimotor resources, effectively orchestrates the maintenance of our joint mechanics and ensures the fullness of our autonomy. The fourth challenge is embodied in the consideration of the elasticity of the mind, both individual and collective. It invokes notions of knowledge, culture and ethics that will be outlined in this book under the strict prism of a voluntarily reductive semantics resulting from the laws of mechanics and elasticity. It introduces the consideration of difference, whether innate or acquired, which opens up the subject to social and societal considerations that respect the values of humans with different elastic potentials, by nature or by accident.

Taking into account the elasticity of the body therefore leads us to manage these four challenges, the management of stress, nutrition, movement and morality. These four challenges become all the more important when technologies seek to exceed the archetypal limits of humanity. The intentions of human augmentation are indeed based on the desire to generate a being with a body that is insensitive to stress, fed in an optimized way, fortified because of technologies that induce tissue regeneration or the replacement of disabilities, and with a mind reinforced by artificial intelligence.

The good news is that it is possible to care for what we will call our elastic capital, by correctly assuming the required choices and actions. Indeed, the proper management of elasticity depends a lot on us, both at the individual and societal levels. We will present what medicine and engineering can do for us, and what we can do for ourselves. By reciprocity, we will explore what the notion of the "elastic human" can bring to society in an interdisciplinary approach ranging from the smallest (genes, proteins) to the largest (the individual and society). Each part can be read independently, relying if necessary on summaries and technical details that can be consulted in the appendices.

In order to present concrete cases, we will recurrently present the situation of people with an accelerated loss of their elastic capital, innate or acquired, i.e., of genetic origin or not. The health news related to pandemics also leads us to introduce the case of people who suffer from the loss of elasticity, sometimes lethal and often chronic, of the lungs or other organs.

As this book ultimately embodies a very personal vision of humanity that cannot be without reference to culture and the spirit of the times, I wanted to introduce seven digital paintings created with mixed media, inspired by famous paintings, as manifestations of elastic thinking.

PART 1

Fantastic Elastic Capital

1

Strengths and Weaknesses of the Elastic Human

1.1. Introduction

The bodies of human beings and animals have been built for and by movement. Those who have been immobilized will surely share my point of view, without denying the amazing faculties of the mind. One can always quibble about the supremacy of the mind over the body, but in reality there is no mind without a heart to feed the brain. A key to this movement is the varying ability of the body's organs and tissues to return to their original form after deformation. This is what defines elasticity. It is what allows the walls of the lungs and arteries to inflate and deflate according to the rhythms of breathing, or the bladder to fill and empty. Addressing the concept of the "elastic human" is therefore a cause rich in implications, as basic as breathing or managing one's movements without untimely urinary and digestive interruptions! The loss of elasticity can induce multiple inconveniences ranging from temporary discomfort to chronic painful syndrome or even, at the extreme, to a partial or total loss of autonomy. More specifically, we can mention heart failure, aneurysms, emphysema, wrinkles, intestinal and lymphatic laziness, ligament rupture, herniated discs and inguinal hernias, rheumatoid arthritis, the difficulties of procreation or sensory deficiencies including the loss of vision and hearing. It is a catalog that is still far too rich to date because it concerns a large proportion of current diseases, chronic or not!

The loss of flexibility and elasticity affects all living beings. Indeed, the elasticity of the body is not renewed or renewed only a little after the

---

For a color version of all the figures in this chapter, see www.iste.co.uk/sommer/elasticity.zip.

completion of growth. This observation leads us to follow the inexorably negative evolution of what must be considered a capital, which wears away during aging, in a visible way in terms of the skin, or collapses during crises, as in the lungs of people suffering from respiratory pathologies.

1.2. Longevity and elasticity

We all have an elastic capital that deteriorates more or less rapidly over time. Fortunately, our life expectancy has increased considerably; it has almost doubled, on average, for populations living in the most industrialized societies. The appearance of wrinkles and elasticity defects is therefore more obvious, or at least we will experience them over a much longer period of time. While this progression of wrinkles is an annoying indicator of aging, the loss of elasticity in all or part of the rest of the body can become more disabling. This leads us to dream of a reprogramming that could modulate the slow decline of our elastic capital. It would be a matter of counteracting the vicious circle that sets in when the elasticity of tissues is reduced, more or less slowly, but always inexorably. Like personalized therapy capable of softening and even saving lives, it would be useful in adults and necessary in children and adults affected by genetic mutations causing a malfunction of elastic tissues. Unfortunately, for the time being, medicine and pharmacology still know little about the causes of this loss of elastic capital because no one knows how to reinduce, replace or repair identical tissues that have become too stiff and inextensible. The reprogramming of body tissues once growth has been achieved remains a challenge for caregivers and researchers. Without going as far as supporting the concept of transhumanism whose objective is to push back the temporal limits of life or to maintain eternal youth, the wish to be able to claim an optimized management of our elastic capital becomes universal.

1.3. Disasters

It is sometimes difficult to measure the inexorable evasion of one's elastic capital when it is hidden in the depths of the body. This state of affairs can lead to diagnostic or therapeutic wandering, for patients looking for an effective management of painful syndromes during their care and life course, when the suffering person does not find any response to his pain. In addition, the appearance of clinical emergencies requires practitioners to find rapidly effective solutions to alleviate suffering and disability, or even to save patients' lives. In the context of current health issues, it should be noted that loss of elasticity is at the heart of the seriousness of several respiratory diseases, such as SARS-CoV-2, which cause exaggerated and sometimes chronic inflammatory reactions that can destroy rather than protect lung tissue. Other pandemics with manifestations and complications associated with loss of elasticity are and will continue to be likely to affect our elastic capital and create an immune imbalance. These considerations, unequivocally, lead us to consider the protection of elastic tissue in the management and prevention of epidemics and endemic or chronic diseases.

1.3.1. *Fibrosis of infectious origin*

My own research topic concerned the analysis of elastic tissue and its alteration. I approached the notion of elasticity by studying the hardening of human tissues that occurs during the formation of scars and/or that occurs during bacterial or parasitic infections. After a first approach centered on the interaction between oral streptococci, caries formation and the development of secondary endocarditis, I became interested in what is called fibrosis, which was studied at the Institut Pasteur in Lyon in the laboratory of fibrosis physiopathology.

In certain infectious situations, tissues change and become rigid under the pressure of a strong synthesis of proteins and fibers, as if the body were building a barrier to defend itself against an aggression. This is the famous fibrosis experienced by some people whose cancer is treated with radiotherapy and who suffer from this side effect. Sometimes, in fact, the body's defense mechanisms go haywire and the natural remedy becomes worse than the disease. This is precisely what happens to the alveoli of the lungs when COVID-19 results in fibrosis that impedes and in the extreme prevents breathing. The lungs lose their elasticity and we simply suffocate. This is respiratory distress. Those who have suffered from pneumonia, bronchitis, emphysema, bronchiolitis, cancer or other lung diseases can feel in their flesh what the slightest respiratory failure means. It is the anguish of the lack of air, something so essential. It is also the pain of expelling carbon dioxide, when the breath is lacking. Suffocation abolishes all plenitude, with the added bonus of fatigue and loss of energy. As the lungs are the gateway to the body, in contact with the external air, they are consequently the target of viruses, bacteria and other microbes, such as SARS-CoV-1 in 2003 and SARS-CoV-2 in 2019, but also the plague, the Spanish flu or tuberculosis. There is no reason for the litany of ailments to be unraveled when air pollution weakens the lungs and the concentration of urban areas facilitates contagion.

1.3.2. *The great pandemics*

Among the great pandemics that have permeated the history of humanity, the plague occupies a place at the top; during the ancient period, one of the first testimonies of a disease close to the plague rests on the writings of Thucydides in an Athens at war against Sparta. The pathology involved in this account was likely a form of typhus of bacterial origin, with pathogens (*rickettsiae*, for the curious) that attack the blood vessels and spread in the body. To stay within the framework of respiratory distress syndromes, pneumonic plague is rather expeditious. The plague *bacillus* (*Yersinia pestis*, again for the curious) invades the lungs where it infects and destroys the tissue of the alveoli, causing death in 2 or 3 days. It does not give the body time to react. The Spanish flu was not to be outdone, caused by an RNA virus like the coronavirus (we will see later on what this corresponds to). Studies suggest that the Spanish flu triggered a strong immune response and then a storm of cytokines in the lungs (like SARS-CoV-2), with destruction of alveolar tissue accompanied by bleeding and subsequent bacterial superinfection. Patients became blue and death followed rapidly as the infected lungs, filled with water, were no longer functional (this is called pulmonary edema).

In COVID-19, there is fibrosis of the lungs with respiratory stress, a side effect which led to the name severe acute respiratory syndrome (SARS). Let us discredit that the virus also induces renal failure, cardiac pathologies, alterations of the vascular wall as well as embolic accidents, chilblains related to vascular problems and intestinal pain. Alteration in olfaction, vision or hearing is not to be discredited, these being consequences of the nervous tissues being attacked. Moreover, the differences in the clinical picture between men and women, between children and adults, and even between people of different geographical origins complicate the management of the disease. Once again, all these pathologies or medical signs can be linked in one way or another to elastic capital.

There are other pandemics that have proven to be catastrophic for our lungs, with the formation of these hard and fibrous tissues that have lost their elastic functionality. An emblematic case is pulmonary tuberculosis, terribly active in the 19th and early 20th centuries and still the first infectious disease in the world today. The last tuberculosis epidemic occurred in France between the two world wars, with a peak in 1930. My two grandmothers died of it, both in their thirties. As for SARS-CoV-2, the microbial agent of tuberculosis (*Mycobacterium tuberculosis*, still for the same curious people and maybe others) nests and proliferates in the lungs where it can induce a strong reaction in the tissues, with an immune, inflammatory response and a reinforcement of the bronchial walls caused by a fibrous scar synthesis. Fortunately, in addition to the improvement of living conditions, the development of antibiotic treatments has made it possible to contain the disease and remove the specter of the dreaded iron lung associated with drastic hygienic measures. In order not to forget the severity of this scourge, archive images show a succession of patients caged for life in their iron lung. Tuberculosis is still rampant and is one of the leading causes of infectious deaths on our planet. And without being a prophet, we can predict that the emergence of strains resistant to all antibiotics heralds a difficult future for the actors of the health system and for the population. Fortunately, I'm talking about a time that people under twenty years old cannot know, that of sanatoriums, a time when it was prescribed not to spit on the ground to avoid spreading tuberculosis.

While the most dramatic effects of organ distress concern pulmonary respiration, other targets are also impacted by COVID-19. The liver is one of these, albeit in a more clinically discrete manner, at least early in the disease. SARS thus joins viral or bacterial infections such as chronic hepatitis (B, C, D, E...), which can also induce fibrosis. A similar mechanism is found in several endemic diseases such as bilharzia (or schistosomiasis). This parasitosis is still called the disease of the pharaohs because it was common in the Egyptian countryside and has been identified in mummies. It is caused by worms that infiltrate the veins of the legs and lay eggs that circulate in the body and block the hepatic vessels. The liver then tries to isolate these egg-laying sites by producing this famous fibrous shell, which causes liver failure. The image of these children and adults with huge swollen bellies is not very exotic, and it is better to avoid wandering in humid areas of Africa or Brazil to avoid catching what constitutes the second parasitic endemic in the world after malaria.

Our beautiful forests are not free of danger either. We can take a walk in the woods, but we should refrain from eating raw wild blueberries, strawberries or raspberries picked on the spot. This is to avoid the fox tapeworm infection, which can be caught by eating berries contaminated by a small tapeworm deposited by a fox or other hosts in our forests. This echinococcosis can lead to a fibrous hardening of the liver that is incompatible with its functioning. Progression is slow, fortunately, but the transformation of the liver into a kind of stone induces an irreversible functional alteration. The fields and forests of almost all the regions of the great European Alpine crescent are frequently contaminated. These precautions should be combined with the precaution of hiking with leg protection to avoid Lyme disease as much as possible.

1.4. Cutaneous elastic capital

Figure 1.1. *Structures and fluids of the elastic human. The intention of this second painting is to present a vision of the elastic human between solid mechanics and fluid mechanics. The figure is inspired by Jean Lurçat's* Chant du Monde, *exhibited in Angers alongside the formidable tapestry of the Apocalypse. The hanging was created as a testimony to the effects of the atomic bomb dropped on Hiroshima. Numerous nuclear tests took place afterwards, leaving residual atmospheric radioactivity, traces of which can be found in the form of radioactive carbon in the biosphere. It is by following this biomarker constitutive of certain amino acids of the elastic fibers that it was possible to estimate the strong stability of these fibers. The analysis was carried out on the lungs of individuals aged 6–74 years who died of various causes during the period 1962–1970, which does not augur well for the longevity of elastic fibers in other tissues with faster turnover, such as the skin or intestinal mucosa (Shapiro et al. 1991)*

1.4.1. Introduction

The questions that my friends and colleagues first ask researchers specializing in the field of elasticity are most often about the skin and its appearance. This often triggers a request for advice on the best anti-aging cosmetic cream, sometimes accompanied by a gesture mimicking the spreading of a product on the cheeks (non-promotional answers are given afterwards, with necessary caution). On the other hand, the questions almost never concern the medicine or food supplements that help improve breathing, digestion or induce more efficient movement. And yet, there is a lot to say or write about. But this isn't what we see in our mirrors, which can be unforgiving first thing in the morning.

1.4.2. Wrinkles and scars

For many people, even though what can be seen on the skin also exists in the depths of our body, talking about elasticity would essentially be a question of the skin, the urgency of which is felt at the appearance of our first wrinkles. Fortunately, wrinkles do not usually indicate skin fragility. The skin always remains waterproof and normally does its job as skin, which is to cover and protect the body. To visualize the problem of maintaining and renewing the skin, we can imagine two scenarios: the first scenario begins with the beautiful skin of a baby and then continues in acceleration by its aging, with the appearance of wrinkles and a skin that dries and withers. There is a disease where this scenario unfolds impressively and at high speed, in just a few years. We will talk about this scenario later, of course. In the second scenario, the skin gets injured, heals poorly, hardens and turns into the hard skin that we see in burn victims. When the wound is small and shallow, the tissue regenerates properly and there is little or no unsightly scarring. After a small wound, the skin regains all its mechanical characteristics. However, when the wound is large or deep, a harder, tauter or puffy area will remain for life. In severe burns, the reformed tissue presents an aspect far from the beautiful properties of flexibility and softness of the original skin. This is called plastic behavior, at the limits of elasticity, when the deformation of the material is no longer reversible when the stress stops. It is this property that has given the generic name of plastic to all these objects that flood the planet.

To return to the skin of burn victims, in addition to the unsightly aspect, the sensation when palpated indicates that the neoformed tissue no longer has the same mechanical properties as the original skin. The skin is taut, hard, stiff and not very flexible. One can imagine tensions, constrictions and certainly ineffable pains. The synthesis of collagen fibers occurs in excess, without elastic fibers. The tissue of some scars is stiff, without the original flexibility, unable to respond to the extensions that are essential for all movements, like a painful spacesuit. We can represent these lines of tension under the effect of different forces which are as many melting pots for the formation of wrinkles. These forces are imprinted on the dermis but also on the epidermis, drawing Langer's lines (with their different layers described more precisely in the Appendix). The drawing of these lines is influenced by different factors: morphology, body type, musculature, age and position of the subject.

The two scenarios described above show two possible developments of the materials that structure the arrangement of skin tissue: aging or the appearance of wrinkles translates the loss of elastic capital while scars and fibroses correspond to the formation of a new plastic tissue or the deformation of an existing tissue. Once again, what is true for the skin is also true for the rest of the body, including the lungs.

There is no truly effective treatment without side effects to induce or optimize skin healing in burn patients or to resolve cutaneous or radiation fibrosis (the latter is a complex side effect of radiotherapy). Only applying pressure in regard to healing or remodeling the skin is truly beneficial. This requires long sessions of compression and extension physiotherapy to encourage the skin to regrow some elastic fibers. This look at scarred skin introduces the notion of a virtuous circle where the application of mechanical pressure promotes the formation of effective and more aesthetically pleasing skin tissue. This local pressure can also be useful in limiting the formation of stretch marks. Here again, we understand all the challenges that lie behind the recovery and maintenance of elasticity.

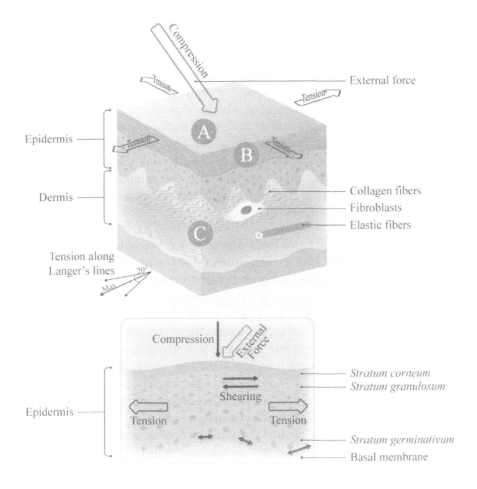

Figure 1.2. Stress transmission in the skin. The diagram illustrates the level and orientation of the forces drawing Langer's lines, at the epidermis (A), the junction between the epidermis and the dermis (B) and the dermis (C)

1.4.3. Beautiful skin

This comparison between healthy tissue and its badly scarred equivalent makes us realize how lucky we are to have beautiful skin (and to feel good about ourselves!). Nature's evolutionary process has selected a technical solution to meet the body's need for elasticity in motion. This solution is different from the one selected by a spider to make its web, a silkworm to make its cocoon, the rubber tree to produce a self-healing latex that humanity would transform into rubber, the mussel to close its two valves or the jellyfish to agitate its gelatinous mass. There are obviously some differences between jellyfish, mussels and humans, but also some common points. Indeed, mussels have their own elastic system, made of different materials than human elastic fibers, but they lead to the same elastic properties. Jellyfish have a system of elastic microfibrils that are not very elaborate and have some similarities with their analogues in humans, which form the basis of the elastic fibers. They also have collagen fibers with a structure close to that of humans. As for the human elastic system, its sophistication has become more complex with the evolution of the heart, primitive versions of which exist in mussels. But not in jellyfish, which do not have a heart as all swimmers know!

The skin is a magnificent tool, not only for seduction but also for mobility and protection. But it must protect itself because it takes the full force of the sun, beneficial for the body but destructive in high doses for skin fibers. The skin first protects itself by producing a pigment that limits the penetration of ultraviolet (UV) rays. This is melanin, produced by the cells of the epidermis that are called melanocytes. These are unfortunately famous because the moles that concentrate them can lead to devastating melanomas. Melanin protects the epidermis against oxidizing molecules generated by UV-irradiated cells. The more melanin the epidermis has, the more effective the protection of the DNA of its cells will be. There are two main types of melanin: those of darker black and brown skins and those of lighter skins. The latter are less effective against the sun's protection and must absolutely avoid prolonged exposure to the sun (or sunbeds with UV rays) which will transform our beautiful skin into indelicate parchment, especially for redheaded skin!

The epidermis is constantly renewing itself, sometimes very quickly. Consider how we peel after sunburn when we forget to protect ourselves. As a general rule, the epidermis reproduces itself almost entirely every 28 days, leaving behind on our bedding and shoulders the dross of dead cells that are our scales. This is a protective mechanism to prevent changes in the genome of cells from becoming toxic under the action of the sun, which may result in cancer. It is partly because of this renewal of the epidermis that scars close quickly to cover wounds. It is also because of this potential that it is possible to reconstruct portions of epidermis in culture, from the cells of the epidermis of those who require skin grafts. In addition to its protective role and this extraordinary capacity for renewal, the epidermis also has a mechanical function by ensuring resistance to stretching. This capacity is maintained throughout life, even though the epidermis becomes less thick with age and its link with the underlying dermis loosens.

For the anecdote, if the cells of the epidermis are replaced, the ink of a tattoo remains present in the skin. The ink is in fact maintained in cells of the immune system which are present under the epidermis and can also renew themselves. The ink then passes from the original cells (called macrophages) to the new cells. This is how an interesting biological property ensures that these skin paintings are maintained for life with ink, which must be chosen with caution. But since we are here, let us continue our investigation on the dermis side.

1.4.4. *The elastic network of the dermis*

It is in the dermis that we will encounter the main protagonists who ensure the beautiful skin mechanics. More specifically, these are the various forms of fibers. The most numerous and strongest are collagen fibers. Their specialists have characterized a whole range of them according to their composition and structure. To give an image, as we have become accustomed to do, we could imagine the ropes of boats at the time of the sailboat. Three or four strands of initial threads, of vegetable origin such as hemp, were wound together and then "committed" again with others in the opposite direction of the winding of the strands. The change of composition, added to the complexity of the committing, provided ropes specific to each use (the name "rope" was only used for the ship's bell). In the same way, singular human collagen fibers are formed by assembling three strands and then intertwining them in a specific way to create a very solid skein, relatively unstretchable, which will structure the dermis and many other parts of the body. This meshwork will be synthesized very quickly during the healing process. But it is also the exaggeration of this meshwork that supports the continuous formation of fibrous tissues that constitute raised scars known as keloids.

At this stage, the elastic fiber comes into play which will provide the skin with the capacity to extend and give it its flexibility. As for collagen fibers, there are several categories. We must imagine them as the weft of a cotton or silk fabric on which a more elastic component is deposited, like a rubber deposit. The substance that gives maximum elasticity to our body is the well-named elastin interwoven with an underlying network of microfibers (called microfibrils). The final weft will become more extensible because of this contribution of elastic materials and will quickly find its starting position after stretching.

The expertise of the complexity of these fibers and their formation must be left to scientists and people concerned with inborn or acquired diseases (Jacob 2006). A diagram of the formation of elastic fibers is presented in Figure 1.3 (with details in the Appendix), from which we will retain at this stage only the notion of a process in several stages (from a millionth of a second to several days or even weeks), involving several tools (DNA, proteins, cells, matrix and structures around the cells).

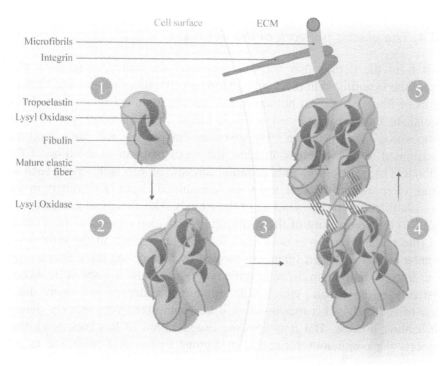

Figure 1.3. *Elastic fiber assembly process. The formation of elastic fibers thus requires the deposition of elastin on a skein external to the cell (microfibrils). An important stage of this assembly is that of its consolidation outside the cell, under the aegis of various molecules including fibulins and lysyl oxidase, the latter being partly affected in the people represented in Figure I.1*

Beyond the composition of these fibers and their formation, the essential thing is understanding their presence and function, at least at the beginning of our lives. This is where the problem lies: these fibers are only really synthesized in sufficient quantities during growth, i.e., up to the age of 20, unlike collagen fibers which are synthesized throughout life. In short, we have a capital of elastic fibers developed in all our tissues during our first years of life, but this stock will diminish with time, as our mirror stubbornly shows us when we reach our thirties. And unfortunately, neither make-up nor plastic surgery will change this state. In addition, we are talking about plastic surgery and not elastic surgery, which has yet to be invented. Many researchers around the world have worked on this subject without achieving satisfactory results, leaving room for treatments that may be aesthetically interesting but that will never reclaim the magical period of youth.

Elasticity potential becomes optimal at the end of the growth phase and then reduces with time. This is not an equitably managed capital, as some people will show almost no wrinkles. Science is currently unable to provide a satisfactory explanation for these discrepancies. The type of skin is surely an essential factor. The presence of black or brown melanin is the most protective, as we have seen. The deleterious effect of exaggerated exposure to UV rays on elastic and collagen fibers is well identified. Their degradation contributes to the appearance of these brown spots poetically called cemetery flowers[1]. These are partly due to the slackening of a kind of membrane that makes the link between the dermis and the epidermis. These lentigines appear with age, especially on skin exposed to the sun which sees this basal membrane weaken and receive concentrations of melanocytes colored by melanin. We can also imagine psychological variables, when serenity and spirituality result in an absence of wrinkles and, conversely, stress leaves its mark and digs into the pockets of our elastic capital.

1.5. Loose skin and cutis laxa

At first glance, beautiful, soft skin is a common characteristic of babies and young children. What a tactile joy it is for their parents to touch these juvenile skins that respond smoothly to pressure and easily regain their silky appearance. But this is not the case for all children. Pathologies due to gene mutations cause malformations of the fibers that structure the tissue. These

---

1 This is a literal translation of the French name *fleurs de cimetière*.

genetic disorders are very rare because the presence of these fibers is actually essential to life. Some mutations can lead to the formation of modified proteins, which become less functional without preventing life, a difficult life but a life nevertheless.

We must now mention the emblematic situation of children with loose and wrinkled skin. This appearance generally indicates the cutis laxa syndrome, where *cutis* refers to skin and *laxa* refers to loose. It is a very rare disease essentially of genetic origin, so rare that only one or two cases were known in France at the beginning of this century and hardly more throughout the world. It took great chance or exceptional circumstances to encounter anyone with this syndrome. My experience with cutis laxa really started on September 12, 2001. The coincidence of the date will not escape anyone; the Twin Towers of New York had just disintegrated, the world was holding its breath and there was talk of an apocalypse. So, when on September 12, in a worrying end-of-the-world situation, I discovered cutis laxa because of a television show, I was impressed by the reality described by the participants. The production had people with skin diseases participating. It was a double whammy, the pathology itself and its visibility to the world. This television format where everyone exposes their problems on screen now seems quite normal, but this was not the case at the beginning of the century. I saw it as a research subject where everything was to be discovered in the field of study that concerned me, both to understand and to help and propose therapeutic solutions.

The participants in that television show on September 12, 2001 presented their experiences of living with non-pigmented spots caused by vitiligo, with very visible scars that do not heal, or with budding acne, until a little girl with big black eyes suffering from cutis laxa syndrome appeared. She must have been about 10 years old at the time, but there was already this destructuring of the skin of the face flowing under the weight of gravity that one normally finds in much, much older people. Her mother explained that her daughter was suffering from an orphan (rare) disease, probably of genetic origin, without being able to name the affected gene. The little girl lived like any other child of her age, notwithstanding her prematurely aged physical appearance. She suffered at times; children can be tough on each other. But her dynamism and her joy of living had embraced all those around her and had opened the doors of camaraderie that for others would have been closed. She captured people's attention and friendliness with her big black eyes and her voice that sounded like she was a heavy smoker. Many questions arose, including how she coped with appearing old when she was actually young, how she managed the various physiological complications accompanying this syndrome and whether the disease would affect the longevity of her life. There were no answers at the time, and they have only been partially answered two decades later.

It was after this television show and an appearance in a local telethon that the little girl with the big black eyes aroused the interest of a group of researchers. A thematic network was initially led by the team of the dermatology department of the Necker Hospital in Paris. The collaboration was then extended to Genethon. The partnership was extended to my CNRS team at the Université de Lyon (Institut de Biologie et Chimie des Protéines) and then to the medical genetics center in Ghent, Belgium. Twenty years later, the Cutis Laxa Internationale created for the occasion made it possible to link several hundred families suffering therapeutic wandering and desperate parents throughout the world. Clinical and genetic studies have allowed the identification of a dozen genes whose mutation is responsible for cutis laxa. We now know that there are different forms that require specific follow-up. Some are dominant mutations, i.e., they override the non-mutated form when it still exists (provided by one of the two parents). Some are recessive mutations which will become visible only if both parents transmit the mutation. The dominant forms of cutis laxa are often associated with a mutation on elastin. It should be noted that the discovery of these mutations has benefited from the progress of sequencing following the almost complete decoding of the human genome in 2004 (it was not considered finalized until 2021). It is good to measure from time to time the progress of science, which in this case has advanced the diagnosis of several very rare diseases, as it has allowed the identification of SARS-CoV-2 and its variants at an unprecedented rate.

As is often the case in biology, the understanding of a complex system benefits from advances in knowledge about its defects. This translates into a shared work between analyses with patients and causality studies carried out on different models, including animals when there is no other solution. The cutis laxa research has worked on clinical observations of affected persons as well as on blood samples and biopsies (collected after information and consent of the concerned persons in the respect of fundamental rights, which is never useful to remember). With very rare skin diseases, it is not always possible to work on the skin of mice that are either too hairy or too hairless. The consortium therefore used a model of skin reconstructed in the laboratory from the skin cells of children suffering from cutis laxa, because of the collaboration with the skin substitute laboratory of the Edouard Herriot Hospital in Lyon (Claus et al. 2008). Fortunately, a few cells are sufficient, obtained from a biopsy necessary for histological identification. Studies have progressively been carried out on mice, some of which had only a limited potential to produce elastic fibers, because of a team from Washington University in Saint-Louis in the United States and another from Grenoble (Fhayli et al. 2020). The zebrafish model, the cute little danio of the aquariums, has also been used.

These interdisciplinary studies, carried out in collaboration with colleagues in France (Lyon, Grenoble, Paris) and abroad (Ghent, Saint-Louis, Pittsburg, Beirut), have taught us a great deal about genes, proteins, mutations, cells and tissues. The starting point of this work is the meeting with families, children and adults, where members of these families suffer from alterations in the formation of elastic and collagen fibers. We have already mentioned children who look older than their age, although their life expectancy seems fortunately quite similar to that of the general population. Encountering a child suffering from progeria syndrome made us feel very strongly that basic research can take too long because this disease, due to a chromosome copy error, grants a life expectancy of less than 20 years. We compared the torsion of our fingers and wrists with those of children and hyperlaxed adults suffering from collagen fiber diseases, such as Ehlers-Danlos syndrome. We envied the pianist's hands of people with elongated limbs and tapered fingers suspected of having a disease called Marfan syndrome, who prided themselves on being like President Lincoln. We were entertained by some merrymakers with mischievous eyes who had Williams-Beuren syndrome, which some call elf syndrome. And then there were the very sensitive encounters with children with stunted or defective growth. Some are on the verge of becoming blind, deaf or disabled and others are sadly no longer with us. A summary of the extreme complexity of these various syndromes is presented in section A.4 of the Appendix and Figure 4.2.

Exchanges with these children or adults are always a source of emotion on both individual and collective levels. Moreover, because of the knowledge of genes and their mutation, scientists, clinicians and associations have supported the genesis of genetic families where emotion exploded at the sight of alter egos, morphologically very similar, like a reflection caused by the same mutation on the same gene. This has considerably helped families and individuals with syndromes who no longer find themselves alone. Because of the Internet and discussion platforms, links have been created and information has flowed around the world between families and medical professionals who have never had the opportunity to encounter such rare diseases. They are indeed very rare because children born with an elasticity defect are in a way survivors since one cannot live without elastic fibers. Whatever the defects involved, they still allow life, even if it is sometimes a question of survival for these new families defined by the disease which is not always of genetic origin. In fact, these last few years have seen an increase in the number of people who have developed cutis laxa without any predisposition or suspicion of a genetic defect. This acquired syndrome thus joins the cohort of acquired chronic diseases with imprecise causes, as are many other diseases where environmental, physiological or autoimmune factors are suspected.

1.6. What is missing and malfunctioning in these fiber diseases?

Research on cutis laxa and other similar genetic diseases has thus advanced knowledge on many pathologies that went far beyond loose skin and whose patients were confronted with therapeutic delay. There are still many orphan diseases for which the precise cause has not been identified. Globally, a 10-year-old child with cutis laxa or Ehlers-Danlos may face the same problems as a 60-year-old person because their elastic or collagen fibers have a construction defect. The mutation of the little girl with the big black eyes concerns the emblematic protein elastin, but many other proteins can be mutated in these syndromes.

Apart from genetic causes, of all the proteins that constitute elastic fibers, it is often elastin that becomes limiting because it is synthesized very little after the growth stage. Once produced, this protein is very stable, perhaps one of the most stable in the body, fortunately since it will not be replaced or will be replaced only slightly. It has been estimated that half of the quantity of elastin produced during growth remains around the age of 70. This estimate was made on the basis of the analysis of tissues of people subjected to radiation after nuclear tests. Although the renewal of this kind of experiment is not really recommended, it is nevertheless true that elastin is one of the most stable proteins in the body. Since we cannot live without elastin, as studies on elastin-deprived mice have shown, a decrease of half every 70 years would lead us to estimate an absolute life expectancy of around 140 years. This is obviously only a simple hypothesis that some people might want to test if they were supplied with beautiful elastic fibers. But well, we will have to wait a little longer and avoid the radiations that destroy elasticity and induce severe fibrosis.

1.7. What structures and strengthens the elastic system

When we speak of what supports the body and allows movement, we also enter a domain that can be associated with death and immobility, the opposite of mobility. In order to explain the mechanisms at work, we must position ourselves on the global scale of structures. What do we actually need to know that will be useful to us later on? When we board a boat, we must start by knowing the names of two or three types of ropes so as not to be caught out by the skipper: rope, halyard, sheet and mooring line are enough to start sailing. Elastic fibers and collagen fibers will be enough to begin to navigate the body's mechanics, although there are other fibers of an extracellular matrix that we will mention in the Appendix. These components intervene to shape the elasticity of living materials, which we evaluate empirically by palpation or mastication, the skin being more elastic than bones and less than tendons and ligaments. The elasticity of these biomaterials is measurable, according to a scale defined by a modulus of elasticity called Young's modulus (see Figure 1.4 and the Appendix). In the body, bones have the highest overall Young's modulus and will appear stiff; they will deform less than ligaments and tendons, which have the best overall elasticity in the body, in any case much more than soft organs like the liver. Stiffness adds the notion of section, a massive bone being stiffer than a thin bone. Lastly, the hardness of the skin and the bone is obviously different, with the surface of the bone offering much more resistance than that of the skin (it is necessary to press much harder to deform bone due to the higher resistance of its surface).

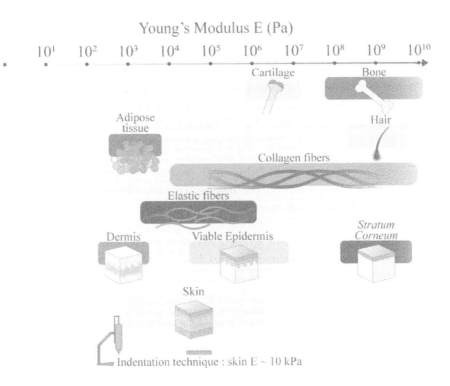

Figure 1.4. *Positioning of organs and tissues according to their Young's modulus*

1.7.1. The mechanical actors

Some will want to place in history, with good reason, some star ingredients in cosmetics, such as the famous hyaluronic acid and some of its affiliations such as proteoglycans. These play on the viscosity, compressibility and lubrication that are easily understood as important elements for the effectiveness of body mechanics. They are used in cosmetic or reconstructive surgery, although their local insertion does not result in an identical functioning to the original. As for elastic fibers, they are not used in surgery because there are simply no natural sources available to produce them at this time.

When performing blood tests, everyone may have encountered fibronectin, which is a sugary protein found in interstitial tissues and also in the blood, where its measurement may indicate a risk of high blood pressure, undernutrition or obesity. Fibronectin should not be confused with fibrinogen, which is a blood coagulation factor that can indicate risks of thrombosis. It is therefore a marker whose excess raises questions about the possibility of blood clots blocking vessels. Once again, the news about COVID-19 and the very small risks associated with vaccination force us to mention this thrombosis mechanism, which is strictly necessary in the first stage of a wound to seal it, but in a proportionate manner.

We cannot talk about mobility and posture without mentioning the muscles which put tension on the tissues. Indeed, elasticity is also one of their characteristics, thanks in particular to a protein called titin or connectin (Linke 2018). While elastin is the most stable protein in the body, titin is the longest and most abundant in striated muscles. It has the ability to extend and regain its shape, providing flexibility first to the muscles and consequently to the limbs. The extension of this protein supports the contraction of the muscle fibers. It must therefore participate in flexibility, although it is difficult to find the demonstration of this hypothesis in the treaties on kinesic practices, in the relatively disappointed hope of making the link between scientific knowledge and physical exercise for all (Alegre-Cebollada 2021).

In this regard, and to finish this list, the role of lipids is essential in tissue formation, but as always, useful, non-oxidized fats are needed. One example of the link between elasticity and lipids is the mucus lining of the walls of the bronchi and the alveoli of the lungs. It is one of the fundamental elements of respiration, preventing the alveoli from collapsing under pressure. This alveolar mucus is rich in cholesterol which is an essential element of cell membranes and the body. We must distinguish between good and bad cholesterol, which are HDL and LDL cholesterols, respectively. Indeed, HDL cholesterol promotes the elimination of excess cholesterol through bile. LDL cholesterol, on the other hand, tends to accumulate in the vessels. But while natural selection has led to the synthesis of these cholesterols, it is because they are good by definition and must be protected and not damaged by excess food or by industrial pollution.

We will finish this biological catalog here, as it will surely frustrate the specialists and perhaps bore the laypeople. However, it is necessary to address a second notion relating to the reinforcement of these structures until they lead to the pathological formation of fibrosis, from which many individuals affected by COVID-19 or other infections have suffered, are suffering and will suffer in the future. This is cross-linking, which corresponds to a kind of networking of these fibers.

1.7.2. Cross-links

It is not lost on anyone that we differ somewhat from jellyfish. Having said that, we should not be complacent, we have seen that jellyfish possess both collagen and microfibrils, the premises and media of our elastic fibers. One of the reasons for our differences with the jellyfish is the possibility of meshing, stiffening and calcification of body fibers. At least three major mechanisms are involved in the formation of this biological network. The first process is intrinsically based on the composition of the fibers, the second deals with the links between the fibers and the third deals with calcification.

The basic structure of tissues is first built by collagen, or more precisely collagens, the most abundant family of proteins in the body. Nature is efficient and has not been overly detailed in this matter of mechanics: collagen is the most abundant protein, elastin the most durable and muscle titin the longest! In reality, about 30 types of collagen can be involved depending on the tissues involved, such as bone or cartilage (Maria et al. 2020). Consequently, genetic diseases associated with collagen defects appear diverse and varied, such as osteogenesis imperfecta syndrome, brittle bone disease or the different versions of Ehlers-Danlos syndrome.

A first program of reinforcement of the structures is ensured by the formation of cross-links between the fibers. While the elementary bricks of the fibers are essential, the cross-linking will reinforce and coordinate the assemblies. We can take the vision of the ropes of the great sailing ships of yesteryear. By analogy, the cross-linking would be an action of winding (commettage) and covering (tarring) of the initial rope in order to provide greater resistance and extension characteristics.

Cross-linking is fundamental for the maturation of elastic fibers. On the other hand, collagen fibers can be built without this cross-linking, but their properties are not optimal and lead to deformities. The consequences of a cross-linking defect are illustrated in the Prado Museum in Madrid. It is necessary to return to the engraving of Francisco Goya describing the effects of the famine during the invasion of the Napoleonic army in Spain. The natives were poisoned by the sweet pea flour that was used to make an ersatz bread. This flour contains a toxin that inhibits the cross-linking of collagen and elastin (as well as a neurotoxin). As a result, their limbs were no longer able to support them (the same story is told in Andrzej Wajda's 1965 film *Popioly*). Conversely, an excess of cross-linking leads to a hard and inelastic tissue. Fibrosis can give a stone rigidity to certain tissues, in an exacerbated form. This could be the case for the liver after ingestion of the tapeworm picked up from eating wild fruits, generating echinococcosis (as mentioned earlier).

The density of the tissue network is an aspect that also concerns cancer. Intuitively, one can understand that tumor cells will not move similarly in hard or soft, plastic or elastic tissue. This is one of the interests of breast palpation to detect if there is a formation of nodules, which are formed by tissue remodeling around the breast ducts. Many scientific teams have been and still are interested in the constitution and structuring of tissues around tumor cells. They have been able to demonstrate how much the proliferation and the fate of the cells were dependent on their mechanical environment within what is called the stroma of tumors (Yamauchi et al. 2018).

This dependence of cells on their mechanical environment has also been demonstrated for blood vessels. Elastic fibers and collagen fibers are thus essential to the functioning of blood vessels that are more or less affected by heartbeats. An experiment was carried out to study the formation of these vessels in the absence of elastin in mice (Li et al. 1998). The researchers expected that mice genetically deprived of elastin would not develop, at least as early as the embryo stage and at least as soon as the heart began to beat *in utero*. These little rodents, totally deprived of elastin, developed to term and only died after delivery, but not by an expected bursting of the walls of the vessels or of the heart, as in a generalized vascular accident doubled by an infarction. The animals actually died because the cells in the blood vessel walls proliferated and eventually clogged the arteries. The cells lost control of their growth and multiplied anarchically, much like cancers. The loss of elasticity (or more broadly the change in mechanical properties of the tissues) thus destabilized the cells which will tend to multiply. This experiment highlights the vicious circle induced by the loss of elastic capital.

I have studied at length one of the mechanisms of cross-linking. It is well known that researchers have their pet causes that appear at the start and accompany their discoveries throughout their professional life. I have personally studied one of the processes that initiates the cross-linking of fibers in fibrosis and elastic tissues. This process is a bit like the catalyst of commercial epoxy glues. It is this catalytic mechanism that is inhibited by the sweet pea toxin. For the curious, the process is activated by enzymes that oxidize lysines, hence the generic name of lysyl oxidase (usually called LOX). Lysines are an important component of fiber proteins.

To summarize, cross-linking is essential for the formation of elastic and collagen fibers. It can be inhibited by diet with a generalized weakening of the body, as in the scene depicted by Goya. This biological mechanism can conversely be exaggerated, which also leads to diseases. Therapies are being studied to target this process in serious diseases such as rheumatic arthritis, but the presence of troublesome side effects hinders their therapeutic use. Cross-linking is also a target for therapeutic research in COVID-19 with regard to the risks of pulmonary fibrosis.

1.7.3. *Calcification*

The other most well-known fibrous tissue reinforcement is calcification. The body contains about 1 kg of calcium located mostly in the collagen fibers of bones and teeth. The rest is necessary for cell and muscle function. The regulation of calcification is very precise: a lack of calcium will result in decalcification, as in osteoporosis, with subsequent fragility; on the other hand, an excess in the wrong place will lead to painful dysfunction; for example, when calcium is fixed on the ligaments of the knee or hand, on the rotator cuff of the shoulder or on the walls of the veins and arteries, which in this last case results in atherosclerosis.

I have personally experienced a severe arthritis attack with the formation of calcium stones in my knees. The pain generated by this gonarthrosis is an ordeal not wished on anyone. However, since the disease is called pseudo gout, it provided an easy and effective angle of mockery for my loved ones when the cause was infectious, likely viral, and had nothing to do with excess food and calories. This confirmed to me how humor (and affection) from loved ones remains a good adjunct to therapeutic practices. Although the laughter is sometimes grating, as when one of my surgeon colleagues ironically gave me an appointment in a few years to cut and "prosthetize" all this. This is an appointment I really want to miss by adopting adapted practices, between sport, kinesis, healthy food and other approaches that will be presented in the second part of this book.

The growth of bones and calcification of neoformed collagen fibers are permanent. Bones are renewed globally every 10 years after growth. This is a long time, much longer than the permanent renewal of immune cells, the weekly regeneration of intestinal mucous membranes or the monthly regeneration of the epidermis. But it is faster and more efficient than the renewal of muscles, not to mention certain neurons of the cerebral cortex and oocytes whose stock is limited in adulthood.

In the case of a bone fracture, fortunately, the speed of repair is accelerated. If you had to wait 10 years after a ski fracture, there would probably be fewer people on the slopes! Bone reconstitution mobilizes the recruitment of stem cells at breaking point. We must insist on this stem cell story because it has become one of the great revolutions in medical engineering. The body indeed has capacities of regeneration hitherto unsuspected. However, this ability is limited to certain organs and situations. We do not have the regenerative power of the salamander (which was thought to have magical powers in the Middle Ages), but the human body is not without the possibility of reconstruction. The most famous case is that of the liver, which can be reconstituted almost entirely, although not quite identically. This power refers to the image of the myth of Prometheus, the Greek god punished and tied to a rock for having given fire to men. His liver was eaten every morning by an eagle. Greek surgeons therefore already knew that the liver was able to reconstitute itself fairly quickly, even if we know today that its regeneration is not identical. But it is almost the only organ which has this ability. However, the surprising regrowth of a phalanx of the hand has been reported in rare children under the age of 6, who had received no treatment after amputation (Illingworth 1974). So do not venture to reproduce this practice (and protect your fingers!). In the meantime, research is progressing, particularly on the regeneration of certain tissues such as cartilage.

One approach addressed in the case of bone is to exploit the capacities of the periosteum to stimulate repair. The periosteum leads to athletes suffering from periostitis when this membrane becomes inflamed, requiring rest and sometimes anti-inflammatory medication. According to clinical observations, the periosteum improves the reconstitution of the bone by a mechanism that surely involves the recruitment of stem cells (Gigante et al. 2001). This kind of "bone skin" is rich in elastic fibers, at least in the youngest patients, which results in a relatively faster repair.

2

Elastic Capital, Air, Water and Other Fluids

Figure 2.1. *"The Big Breath"*. The third painting pays homage to the necessity of clean air in the elastic human's world. As an introduction, I used the representations of Odilon Redon with his strange ocular representations floating in the air illustrated here in the form of a balloon. We will see that the original balloon is a duly elastic fabric. The lightness confronting gravity is illustrated by the windy hair of the dancers from the ballet Gravity choreographed by Angelin Preljocaj. I chose to paint extemporaneously where the dancers move sitting down, a nod to all those with mobility issues but who still preserve a certain agility

For a color version of all the figures in this chapter, see www.iste.co.uk/sommer/elasticity.zip.

2.1. Introduction

Our body is the incessant object of air and liquid flows; nothing is dry and nothing is empty (except the pleura whose negative pressure maintains the lungs in depression). Fluids circulate and infiltrate in the smallest interstices of the tissues and organs according to routes that owe nothing to chance. Fluids carry fuel and handle the logistics of storage and waste management. They are in trouble when fats and pollutants accumulate and obstruct their routes. They need elastic pumps and hoses that can handle the pressure pulsations and the pressure of gravity and the atmosphere above us. Without elasticity, there is no blood or lymph that circulates, no nerves that electrify, no heart that pumps, no lungs that inhale and exhale, no bladder that holds back, no intestines that stir and expel and no liver that cleans, synthesizes and stores.

So let us look at the logistics of it all, exploring some simple practices, like doing yoga, sipping beer or fixing laddered pantyhose with nail polish. We will talk about anabolism for the synthesis of collagen and elastic fibers and catabolism for the production of energy. For those who are resistant to scientific language, anabolism is a bit like Yang and catabolism like Yin, if we can give Western definitions to these notions of Yin/Yang (Sun et al. 2020).

2.2. Respiration

Life is movement and energy. Air provides the primal oxygen necessary for the production of energy in the body. We need a permanent supply of oxygen (in reality, dioxygen or O_2) since it cannot be stored. In return, we need to eliminate the combustion product of this oxygen in the form of carbon dioxide (CO_2). This sounds very simple, and it would be if the body had time. If we moved little (very little!), the law of gas diffusion would be enough (maybe) to ingest oxygen and release CO_2. To illustrate this law of gas diffusion, it is what causes the smell of French fries to escape from the kitchen to reach the other end of the apartment, i.e., from the most concentrated to the least concentrated. The oxygen of the air can thus theoretically be introduced spontaneously into the body since it is approximately four times less concentrated there. Carbon dioxide, on the other hand, can escape spontaneously because it is more concentrated in air, except if we breathe at the end of a combustion system exhaust or in a very polluted atmosphere. The permanent efficiency of respiration therefore requires a complex and very finely regulated apparatus where the mechanics of the tissues are dominant. This is what we will see in the following paragraphs and in a synthetic form in section A.3 of the Appendix.

2.2.1. *The breathing cycle*

For the gases to diffuse, a good entry point must be present, such as the lungs. Because contrary to that of the frog, the human skin does not allow the subcutaneous passage of oxygen, in spite of its important surface (the skin can have a surface of approximately 2 m^2 and a weight of 4–5 kg). In a way this is fortunate; otherwise, we would be all slimy, covered with dermal mucus to protect the skin. We are still able to release a little bit of carbon dioxide through the skin, but less than 1% of the total, which is not enough to turn us into a frog!

Human breathing involves first the lungs, which themselves have a sticky wall because they are covered with mucus, at the base of the formation of our delicate mucus. Oxygen from the air enters the body through the trachea and bronchial tubes with their relatively elastic walls. Then, it reaches the pulmonary alveoli and crosses their walls to reach the veins located in the vicinity. The space between the walls of the alveoli and the veins is very thin and filled with elastic fibers that allow the cavities to inflate and retract rapidly.

This mechanism is very nice, but it would be far too slow if there was no agitation process. It is indeed necessary to dissolve quickly the oxygen in the blood, thanks in particular to blood circulation. This principle of dilution is the same as that of mixing sugar in coffee; the sugar can diffuse by itself, but it takes its time and the spoon pushes it to mix quickly. Oxygen is therefore taken on board by the pulsed blood where it is effectively mixed double-quick. It is then carried to all the corners of the body where it will be released in the tissues, organs and cells which ensure the formation of energy.

The combustion of oxygen is accompanied by the formation of waste, just as a wood fire produces smoke and ashes. Carbon dioxide is the main waste that the body takes care of since it eliminates what it produces (unlike our current societies that are satisfied to store toxic waste for thousands of years). Carbon dioxide leaves the cells that produced it, goes through the veins, then through the lungs to be sent into the air, where it is less concentrated. It is like the smell of French fries coming back from the end of the apartment to the kitchen if the window has been opened (remember... always from the most to the least concentrated). In addition, the cardiac system assumes the propulsion of blood, with its two halves of the heart, the left side that sends oxygenated blood to the organs that need it and the right that pushes the carbon dioxide to the lungs where it diffuses to the outside. This distinguishes us from jellyfish which have neither heart nor lungs, and yet absorb oxygen by the simple law of gas diffusion.

Of course, elasticity is needed to keep all these beautiful mechanisms well oiled, or more precisely to allow the alveoli of the lungs to inflate like balloons (Figure 2.2). The 300 million alveoli lined with elastic fibers could cover a tennis court of about 75 m^2 all curled up between our ribs, which appears to be quite damaged by numerous viruses, pollution or cigarettes (Boraldi et al. 2022). The example of the laddered pantyhose allows us to understand the risks of damaging this lung tissue. Pantyhose (or stockings, made of silk or synthetic fabric) are stretchy so that they fit the movement of the legs. When pantyhose are laddered, the stretch is much wider in the areas where the mesh is disorganized. The fabric does stretch more, and it deforms under pressure, but it has a hard time returning to its original shape. We speak of an increase in compliance which means an increase in the possibility of extension but a loss of elasticity. This is what happens with emphysema when the network of fibers is degraded, and the alveoli widen more easily in places during inhalation but do not restructure well during expiration. It then takes much more air volume to inhale oxygen and especially the air loaded with carbon dioxide is more difficult to exhale. This is often the case for cutis laxa children. We will also find this kind of phenomenon in abdominal hernias.

One way to limit the threading of the pantyhose mesh worsening is to treat the affected area with lacquer or nail polish. This limits the problem, but it still results in stiffening and a decrease in the elasticity of the pantyhose. This is what happens on the sites of pulmonary fibrosis, where the disorganized pile of collagen fibers tries to protect or repair the alveoli's supporting tissue. Natural oxygenation becomes less effective and in extreme cases requires oxygenation on an artificial ventilator. We can therefore say that emphysema and pulmonary fibrosis result from different modifications, the first with less elasticity and the other with more rigidity. Inhalation and especially expiration are much more difficult.

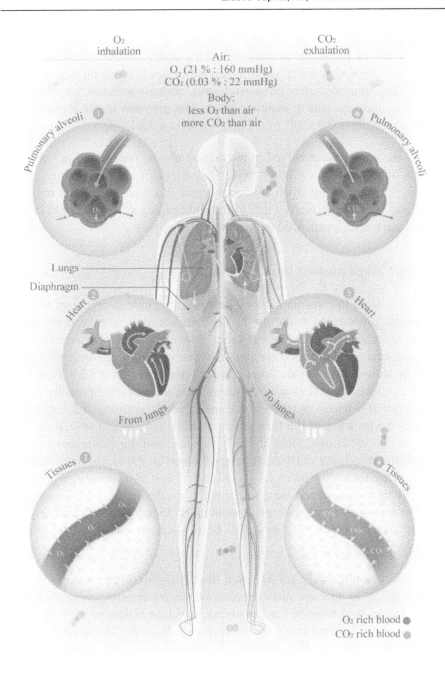

Figure 2.2. *The elasticity of the body and the breathing cycle*

In parallel, elasticity is also needed to enable the proper functioning of the heart valves or in the arteries and veins to absorb the heartbeat. It is necessary to have an association with driving forces to move all this, such as the cardiac muscles, the intercostal and abdominal muscles and the pericardial muscles. Not to mention mucus that prevents the collapse of the alveoli and protects the walls of the lungs and bronchial tubes.

This elastic system involved in respiration can be considered as being at the peak of mechanical evolution, with its emblematic elastin appearing late, probably with the mammalian cardiac system. But this mechanism remains fragile and must be perfectly oiled in the right places and at the right times. We will see which levels need to be adjusted!

2.3. The 12/8 of perfect blood pressure!

The first level to consider is blood pressure. It is indeed necessary to have an efficient pressure to make the blood circulate since the heart moves a mass of blood equal to about three times its body weight per hour. But too much pressure and the device may burst, not enough pressure and everything works in slow motion. This translates into the sacrosanct 12/8 figure of optimal pressure. The number 12 corresponds to the pulsating pressure of the blood when the heart contracts and expels blood (systole), the number 8 measures the pressure of the flow when the heart relaxes and fills (diastole). Nature did not determine these values by chance, and they correspond to the pressure that the heart must exert to send blood to the head or to make it flow from the feet to the heart, all this in the context of atmospheric pressure (divers know that it is even more important when the pressure of the water is added to that of the atmosphere). It is as much to say that it takes a higher pressure and a very big and powerful heart to ensure the same process in giraffes.

As nature is not overly centralizing, the tissues involved have a personal elastic capital to manage their needs. However, arteries, veins, cardiac tissues, valves, and bronchi harden, shrink, and lose elasticity with age, and this is when dysfunction occurs (Cocciolone et al. 2018). The heart manages the deficit by increasing the pumping and pulsing work to inject the amount of oxygenated blood required into the tissues. This allows the body to achieve a high vascular pressure. This fatigues the heart and increases the risk of vascular rupture. This is often the case for children with cutis laxa who see their clinical picture completed by both heart failure and pulmonary insufficiency.

There are two moments when the delivery of oxygen and carbon dioxide can be in trouble, causing permanent discomfort to their hosts. Our precious gaseous acolytes are like those sailors who progress from Charybdis to Scylla, those epic guardians of the Strait of Messina who watch over the west coast of Italy. The first test is in the passage between the walls of the lung alveoli and the vessels just below. The gap is very narrow, and there is usually no problem with diffusion unless there is a change in the alveoli's walls, underlying blood capillaries and surrounding tissues, or if fluid invades the gap. SARS-CoV-2 and other pathogens can lead to this, as can fluid invasion in pulmonary edema related to heart failure. The second event occurs when both gases migrate between vessels and cells in body tissues. The distance to be covered is large on a molecular scale. Navigation hazards can then pile up with fatty deposits or inflammatory areas. Obesity increases the fluidic difficulties and induces a decrease in energy production.

Another danger, besides the absence of oxygen, is an inadequate concentration of carbon dioxide in the air. Its excess prevents exhalation by the lungs of this gas toward the air where it is already too concentrated. This is asphyxiation, either rapid in front of an exhaust pipe or slow and chronic in the atmosphere of our polluted cities. Nature has understood the priority for internally measuring carbon dioxide because it generates the breathing reflexes because of the presence of sensors that continuously measure the carbon dioxide content of the blood rather than that of the oxygen.

These scientific considerations are translated in very concrete terms, as in the yogic breathing for example (pranayama). Let us add that ventilation must be optimal by mobilizing the diaphragm which, let us recall, goes down! This is the famous belly breathing, therefore "from below". To avoid inflating the thorax and losing respiratory efficiency, we must imagine the thoracic cage as a loose spring suspended from above and stretched downward by a motor, in this case the diaphragm. The coils at the bottom of the spring stretch more easily, which is the case for the floating ribs at the bottom of the rib cage, which are less constrained because of their greater elasticity. The result is a better performance by opening the lower ribs and optimizing oxygen absorption, which great singers know well. Ventral breathing is seen in children, who lose this good habit later on, who knows why.

To schematize, the composition and organization of elastic tissues are different along the blood circulation chain (Figure 2.3). The large arteries, which are subject to the high pressures of the heartbeat, such as the aorta, have thick and very elastic walls with a reduced lumen, and allow vascular conduction; the arterioles have a less thick and less elastic wall, but they are contractile to regulate the flow by vasomotricity, which is called vascular resistance; the capillaries have thin walls with a reduced lumen and are not very contractile to promote the diffusion of gases; the veins have thin, deformable and not very elastic walls, with a large lumen, which facilitates the appearance of varicose veins. Peripheral vessels irrigate the entire body and support the nerves.

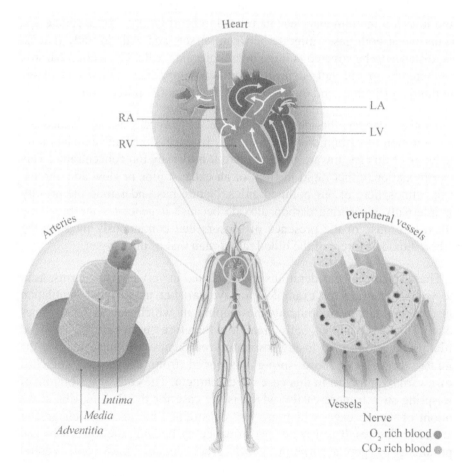

Figure 2.3. *The cardiovascular elastic system*

2.4. Cellular respiration and energy

Here, we come to the crucial moment when oxygen is made available to cells to produce energy. Most of this production takes place in small oblong boxes, the mitochondria. At first sight, this has little to do with elasticity, but we cannot talk about body mechanics without mentioning energy, its production, storage and delivery to the sites of tissue action. This explanation will allow us to say a few words about those situations we hear about without really knowing where they are, such as aerobic, anaerobic (lactic) and ketonic breathing, which all contribute to the formation and release of energy.

2.4.1. *ATP, our universal source of energy*

The body needs energy to function. It must generate energy at all times and in all places. This is catabolism and some will see it as Yin. Food and oxygenation are the primary sources. The universal intermediate storage element is in the form of a phosphate derivative, regardless of the primary source. We must therefore speak of this derivative, which is called adenosine triphosphate (or ATP), because it is our oil, our gas, our enriched uranium or our on-board electric battery! ATP is made of phosphate and a sugar (ribose) attached to a nitrogenous molecule (adenine). Its components are generally abundant in food. These molecules are at the foundation of the formation of the genetic heritage of all living species since they intervene in the creation of the DNA and RNA as well as in the formation of many proteins. Once again, nature plays it safe and uses basic molecules to ensure essential functions, such as providing energy and reproducing the genetic heritage. As for phosphate, it is involved in the construction of bones and teeth, coupled with calcium.

As mentioned, ATP is formed in cells, primarily in a microcompartment called the mitochondria (Figure 2.4). Its most efficient manufacture combines the use of glucose and oxygen. This is called aerobic respiration. If there is no more oxygen available, anaerobic respiration takes over, as it does in the muscles after increased effort. This breathing is much less efficient and cannot be maintained for long, as those who are subject to muscle cramps know. When glucose is hardly available, the body can produce ATP in small quantities using fatty acids or proteins as resources, again less efficiently than aerobic respiration, which should be preferred in all cases when possible.

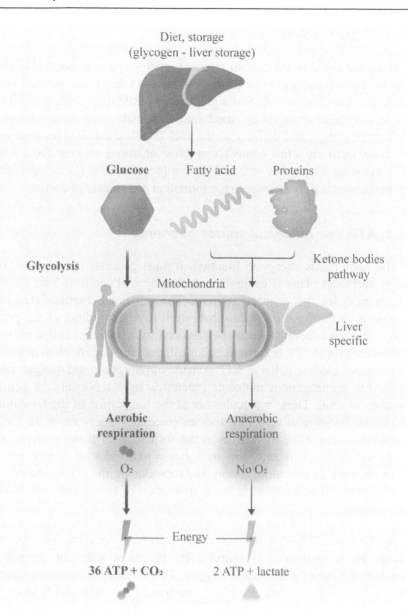

Figure 2.4. *The ATP formation cycle*

Our energy therefore comes from ATP which must be constantly supplied and which will release its energy on demand. This is a limiting element of the process that conditions our quality of life because the body's ATP reserves are low. Oxygen and a source of carbonaceous substances continuously supplied by food or by the removal of body fats are therefore required at all times. The increase in oxygen supply and good air quality will therefore have an immediate effect on energy. So, we breathe deeply as soon as we can, if possible far from pollution, and we eat properly. Those who live at home will find a time to move around, but in the meantime, they will note with interest that magnesium stabilizes the intermediate and final forms of ATP, which can provide them with an alibi to eat chocolate, without excess of course (Maguire et al. 2018).

2.4.2. Our mitochondrial battery

Small structures called mitochondria house the production of ATP. Their malfunctioning therefore manifests itself by a low level of energy production, as in certain genetic diseases. Energy-intensive tissues and organs are particularly affected, such as the muscles in the different versions of mitochondrial myopathy or for the functioning of nervous tissue (including the brain). The functionality of mitochondria is therefore studied in neuromuscular diseases like Parkinson's disease, Alzheimer's disease, chronic fatigue syndrome, fibromyalgia and muscular dystrophy. There is also a theory of aging which considers that the degradation of the mitochondrial DNA accelerates cellular aging and that it is a risk factor in certain pathologies such as COVID-19 (Moreno Fernández-Ayala et al. 2020). There are still many debates about the danger and adverse effects of the use of fungicides, pesticides and herbicides in high doses in agriculture (several substances are suspected or evoked like chlordecone, lindane, the inhibitors of succinate dehydrogenase, glyphosate) or to the risks linked to the excess of heavy metals in food and medicines (cadmium, mercury, copper, etc.). Fortunately, some damage can be mitigated; in fact, the more muscle mass is mobilized, the more efficient the mitochondria will be. There is no mystery about it; moving, practicing sports, oxygenating and eating properly increase our energy.

We dare to end with a little digression on these mitochondria to set the record straight on the distribution of parents' genetic heritage. The cause is understood; the genetic heritage differs between a woman and a man in terms of the X and Y chromosomes. But apart from that, we can assume that the two sexes are equal with regard to the transmission of their genetic heritage. This is true, but only in part. Because there is an exception due to the presence of DNA in the mitochondria that only comes from the mother. In other words, it is this organelle brought by the mother that will mostly generate the molecule that serves as universal energy for the body. Let us be thankful, gentlemen, that energy is transmitted by women!

2.5. The logistics of digestion

Respiration allows oxygen to be transported to the sites of energy formation. In other words, everywhere in the body! To truly use this energy, we need to get the carbon source to the same places as the oxygen and at the same time. First and last kilometer logistics enable the inputs to be routed and managing their waste is not often described. It should be emphasized or recalled that everything takes place in a liquid environment, with the consequent consideration of flows and waves directed by channels, valves (sphincters) and pumps. We have already seen that blood circulation is maintained under the pressure of the cardiac driving force to which we must add peripheral vascular and ventilatory forces, without forgetting the dynamics of the digestive transit.

2.5.1. *The intestinal walls: a history of pushing*

Food is too often chewed in a hurry, as we eat "on the go". It goes down because of gravity, but you still have to push a little. The waves of muscular contraction around the walls of the digestive tract activate food transit, helped by some anti-reflux valves because we are not ruminants! The food, duly segmented and degraded by acidity and specific enzymes, will then be stirred and filtered to be directed either towards the blood in the small intestine or to the large intestine to extract all the juice. The sugars and the majority of proteins will continue their journey through the bloodstream, while fatty acids will join the lymph to enter the bloodstream at the opening of the heart. Unfortunately for those who eat fried foods and other saturated fats, this will spread greedily along the walls of the ventricles and heart valves.

These flows require an elastic apparatus under the muscular walls of the entire digestive system. A special mention can be given to the small intestine, our sorting yard for food products. Numerous folds in its wall, called intestinal villi, make it possible to multiply its surface to facilitate exchanges between the food bolus and the underlying blood or lymphatic capillaries. Below these folds, the presence of muscular layers and elastic tissues will reinforce the wall and will allow food packages to be transported. At this location, abnormal pockets may form causing painful diverticula, sometimes accompanied by the breakdown of peripheral elastic tissues (Ludeman et al. 2002).

Humans have taken advantage of these elastic properties of the intestinal walls by developing the use of balloons, which come from the large intestine of cattle or sheep. These bladders have been used to protect a wound but also to try to protect against unwanted births; this is how we go from the *peau divine* (an archaic French term for a condom that literally means "divine skin") to the "English ridingcoat" (a term used on both sides of the English Channel).

2.5.2. *The lazy lymph*

There is no great driving force in the lymphatic system. This may seem like an oversight by nature, considering its importance, since the lymph has to carry fatty acids, expel waste and carry immune defenses. Fortunately, these lymphatic channels make up for it by having some elasticity and a host of elastic anti-reflux valves (Munn 2015). This was needed to keep this oh-so-slow, lymphatic flow moving without a pump! But a defect in this mechanism quickly leads to a slowing down or even a stagnation of the flow that people with heavy legs and painful edemas experience. Once again, one of the remedies is to move (without shuffling!) because external and internal movements accelerate the flow. Massages are recommended, taking care not to massage the anti-reflux valves the wrong way. Edema is common in people suffering from cutis laxa or similar syndromes, but also in sedentary individuals. Lymphatic dysfunction thus joins the list of indicators of our general condition. One of the testimonies of this dysfunction can be seen in the dark circles beneath our eyes where the very thin skin does not mask the slowness of the vascular and lymphatic flows.

2.5.3. *The kidneys: a story of water and blood*

Abnormal fluid flows can be impressive, and careful regulation is needed to control them. It is the kidneys that play this administrative role with nearly 200 L of blood filtered and reinjected into the cardiovascular system per day and 1–2 L of urine produced. The main function of the kidneys is to filter the blood and then redistribute it. An example will help explain the complex and essential role of the kidneys. To do this, we will evoke the work heavy beer drinkers expect their kidneys to do. Let us transport ourselves mentally to Oktoberfest in Munich where the quantities ingested sometimes defy comprehension. After a first sip of the hopped liquid travels from the mouth to the stomach and then to the small intestine and finally to the blood, the kidneys react directly to the increase of liquid and matter. They do what they know how to do; they send some of the water to the bladder and filter the rest back into the blood. But you do not want to inject too much water because you could get high blood pressure. Like a torrent in a storm, the more liquid there is, the greater the volume and the higher the pressure. The kidneys will therefore have to regulate the blood pressure by calibrating the volumes reinjected. At the same time, these bean-shaped organs manage the quantity of salts. It is a question of respecting primal salinity, inherited from our great marine ancestors. The kidneys also have to manage the quantity of proteins and other solutes in the blood. For heavy beer drinkers, alcohol and the concentration of saline and other solutes will alter the filtration and return of fluids to the blood, leading to the bladder being filled rather than blood volume being addressed. The drinker will be thirsty because he will lack water despite the amount of beer swallowed. At the same time, he will urinate considerably. Note that the history of beer drinkers is similar to that of drinkers of other alcohols, despite what some local distinctive identities say (I will let you answer the question I was asked in Marseille about the effect of the quantity of water to be added to pastis liqueur).

To do their work, the kidneys have at their disposal an extensible and elastic set of highly structured filters, according to a formula already seen in the lungs and the small intestine. The kidneys deploy millions of small structures, called nephrons, which form up to 80 km of tubes. These expandable tubes have a high presence of fibers and elastic fibers whose degradation necessarily has an impact on their efficiency.

In short, the kidneys manage the volume of water in the blood and urine, the concentration of salt (with the change in temperature), the elimination of toxins and toxic products such as pesticides and the stimulation of red blood cells by erythropoietin (EPO is well known to some unscrupulous athletes). Do excuse me, and I am forgetting some of them voluntarily so as not to devote all of this chapter to their functions! There are so many actions that are altered in children and adults suffering from cutis laxa or other symptoms that are very common in people suffering from obesity or diabetes.

2.6. Vascular dilation and constriction

Two mechanisms dependent on the elasticity of the vessel walls manage blood flow, by opposing actions. Vasodilation by enlarging the diameter of the vessels increases blood flow. The veins relax and dilate, their diameter widens, the blood flows more peacefully and blood pressure decreases. This is particularly the case for the veins of the lower limbs, which are larger and offer more time to flow. The induction of vasodilation is a sought-after mode of action of antihypertensive drugs. The body has its tools to induce vasodilation when needed, during inflammation and sports activity, or to dissipate excess heat by an influx of blood to the skin. The nitrogen in the air, combined with oxygen, especially during inhalation, has the ability to facilitate circulation. Alcohol has the same positive effect on dilation, with a feeling of warmth, even when it is cold. This feeling is quickly forgotten because it is immediately followed by a constriction in the extremities so as not to dissipate the heat needed by the central organs.

Speaking of extremity, pharmaceutical specialties based on nitrate derivatives have the ability to increase vasodilation and subsequent blood flow to the erectile functions of the penis, which, among other things, helps to reduce the effects of age and obesity which induce erectile insufficiency (Mitidieri et al. 2020). Certain physiological similarities between the penis and the clitoris have led to the exploration in women of the effect of this active ingredient, which is widely used by men with erectile dysfunction (scientifically, these are phosphodiesterase inhibitors). The hypothesis is to obtain a swelling of the clitoris that can be translated into sexual excitement. Preliminary studies are controversial and draw attention to potential side effects (Bianchi-Demicheli and De Ziegler 2005). So much for the mechanics of desire where love also plays its role. These mechanics lead us now straight to the notion of constriction which is, as dictionaries say, a circular pressure that decreases the diameter of an organ!

As any action implies a reaction, vasodilation may be followed by vasoconstriction. The latter causes an increase in blood pressure since the diameter of the vessels is restricted. Vasoconstriction is induced by an excitation of the sympathetic (or orthosympathetic) nervous system. It is this system that makes the heart beat at full speed and gives us a rush of adrenaline in case of fear or love at first sight. It is through it that coffee has its exciting effect, at least in part. This narrowing of the diameter of the vessels will cause the whitening of hands and feet exposed to the cold. As for vasodilation, it is stimulated by the parasympathetic system, of which the vagus nerve is the best known. It can sometimes be so calming that it can block the heart and make us faint, during a vagal shock. In a way, we could summarize the beneficial actions we offer our body by this gymnastics of our cardiovascular system, between dilation and constriction.

Our actions can therefore have a positive or negative impact on our blood pressure regulation mechanisms. Sport and chronic stress lead to an increase in blood pressure as a result of accelerated breathing. Conversely, the calming effect generated by gentle practices such as singing or yoga is felt by a dilation of the vessels inducing a decrease in the heart rate. However, if we manage to practice a kind of Pilates method of the blood vessels by playing on the dilation and constriction, the coronaviruses are also capable of it. This deserves some attention because one of the blood pressure regulating components is the gateway to SARS-CoV-2.

2.6.1. *The SARS-CoV-2 gateway and blood pressure*

The SARS-CoV-2 virus is malignant, meaning that it hurts us because it has been largely selected to do so. The virus enters the cells of the human body through a doorway that is found on the cells of many organs, such as the lungs, the kidneys and intestines, as well as on the walls of blood vessels or nasal tissue. The singularity of SARS-CoV-1 (the 2003 virus) and 2 (the 2020 virus) is that they enter infected cells through a "brick" that helps regulate blood pressure. By attaching itself to this essential blood pressure control protein, the virus binds to the same targets that nature has developed and that the pharmaceutical industry has appropriately derived to treat hypertension. It is the key point that participates in the regulation of blood circulation in the body. Under the guise of a name that sounds like a Star Wars robot, the ACE2 protein hides a very elaborate mechanism that is essential to our life, and also to that of the virus (Scialo et al. 2020).

Several organs cooperate to manage blood pressure. In nature, we play together. The kidneys are the bosses that receive help from the liver, the adrenal glands, the sympathetic system and the pituitary gland. All together, they act on the tension of the vessels, which is called *angio* in Greek. Hence, the name renin-angiotensin system is given to this regulatory system, which controls both hypotension and hypertension, according to the requirements. Some anti-hypertensive drugs paradoxically target the natural regulators of hypotension. However, SARS-CoV-2 targets one of the steps in the natural regulation of hypertension (through ACE2, the acronym of the offending enzyme, angiotensin converting enzyme 2). When everything is working, that is to say with no virus and with an intact elastic capital, our kidneys and their troops will control the contraction or dilation, the reabsorption of water by the kidneys, and the amount of salt and proteins to be kept or sent to the urine in order to never run out of water. This will not be the case for our thirsty beer drinker, as we have already mentioned.

The unresolved question to date is whether there is a combination of these criteria, between susceptibility to infection, ACE2 content and loss of our elastic capital. This is a very topical question as times are difficult for people suffering from chronic diseases following infections. Without waiting for the answers that will come from future research, it is a matter of including the good management of our elastic system in the arsenal of our life choices and our adapted behaviors, individual or societal.

2.7. Sugar logistics

2.7.1. *The liver and blood sugar*

We have mentioned the importance of sugars for energy production. The maintenance of a sufficient level of sugar in the blood is therefore essential, in the same way as oxygen, since it is the body's main fuel. But unlike oxygen, the body must adapt to a variable rate of sugar intake, which is a function of food intake. A major player in the regulation of blood sugar levels is the liver. When sugar is in excess, it is the liver that will react in the first instance by blocking its synthesis and by inducing the storage of circulating sugar in the form of glycogen or fat. This is the principle of fatty liver production in migratory birds undergoing forced feeding, thus taking advantage of their storage capacity for a long journey. There will be no reference to the suffering of human livers under the pressure of junk food and sugary drinks, as witnessed by the ever-increasing soda disease (the non-alcoholic fatty liver disease that is so difficult to treat) that is added to the excess of fat (fatty liver disease) (Maurice and Manousou 2018). And if, in addition, alcohol or hepatitis are involved, the mixture will be explosive and may evolve toward cirrhosis with a fine fibrosis blocking the irrigation tracts of their liver and, as a bonus, the possibility of cancer. The force-feeding of geese has been extended to a sugar pandemic that is causing havoc!

The role of insulin cannot be ignored at this stage. It is its role as the maestro of blood sugar regulation that makes it known worldwide. Insulin produced by the pancreas acts by lowering the concentration of sugar in the blood, either by encouraging its storage or by promoting its use by the cells. Not enough insulin and the body may not properly regulate circulating sugar levels (which can lead to hypoglycaemia); too much insulin (hyperglycemia) and the excess of sugar in the blood will lead to severe consequences which we will discuss next.

Insulin is part of a family of important hormones. In this family, we find the insulin-like growth factor or IGF1. This hormone is the weaponized arm of the growth hormone. IGF1 stimulates the growth of the conjugation cartilage of long bones, and of all tissues in fact. It improves calcium absorption and glomerular filtration in the kidney. In reality, IGF1 seems to have nothing but qualities to answer all the demands already discussed in this book, especially since it is the only hormone with a documented positive action on the synthesis of elastic fibers. What happiness, I should have told you about it a long time ago, shouldn't I? But it is a happiness that has its counterparts because an excess of IGF1 and other hormones of this family stimulates the formation of tumors, while a low level seems to be correlated with a risk of cardiovascular diseases. Stop the applause and use this hormone only sparingly, as should have been done for growth hormones, whose scandalous preparation resulted in the development of the deadly Creutzfeldt-Jakob disease.

2.7.2. *The kidneys and liver under sugar pressure*

It is time to present the deleterious effect of sugar when it is in excess. I could have talked about it before, but first I had to talk about the liver, the kidneys and the blood vessels, which are the first to suffer. Indeed, excess sugar modifies the constituents of the tissues by promoting a toxic cross-linking. For the sake of clarity, I will call this modification the caramelization of the body, whereas the scientific term is glycation (Fournet et al. 2018).

Let us develop this analogy with the formation of caramel, to illustrate the point. To make caramel, we can start from crystallized sugar and heat it. It then forms this brownish material with different properties giving the famous caramel taste. It is not the same sugar, in reality, but a matter condensed differently that does not have the same properties. In the body, the same thing happens, more slowly of course than in the pan, but just as effectively. The sugar can thus condense and lose its energetic properties. But above all, it can bind to tissue proteins and induce the formation of what specialists aptly call advanced glycation end products (AGEs). This process induces the abnormal cross-linking of many proteins and affects the formation of collagen and elastic fibers. Among other things, in the worst-case scenario, diabetes can lead to blindness, either by "caramelization" of the blood vessels that feed the retina or by the "caramelization" of the very elastic membrane, which underlies the retina and is called Bruch's membrane. This is a well known fear among patients with diabetes. Obviously, what applies to the eyes also applies to the rest of the body, for example, to the walls of the blood vessels, the liver, the kidneys or the lungs.

2.8. The perineal set and terminal delivery logistics

We have seen the need for good elasticity to optimize the logistics of the delivery of gases and fluids that concern the major actors of metabolism, on both sides of energy production, the building up (anabolism) and the breaking down (catabolism). Another form of logistics concerns all the organs stored between the thoracic cage and the perineal space. It can be considered the logistics of terminal delivery of liquids, quasi-solid waste and indelicate gases. It is also the logistics of reproduction, for both sexual activity and childbirth. We are dealing here with organs that are hollow for the most part, especially in women, and are therefore susceptible to crushing or collapsing, or even curling up. They are soft, non-calcified and supported by the peritoneal base without being protected by the equivalent of a rib cage or a skull. Fortunately, they benefit from the support of muscles and the entanglement of fascias consisting mainly of collagen fibers and a few elastic fibers.

2.8.1. *Cross-linking and vaginal prolapse*

Genital prolapse is a reality for many women, and menopause does not help. It is better known under the distressing name of organ descent, which corresponds to the descent of the pelvic organs to varying degrees. For example, a bad cross-linking of the elastic fibers proved to be very serious for mice whose vagina pointed like a balloon outside the animal (Liu et al. 2006). The enzyme targeted in the trail was involved in the cross-linking of elastic fibers that we have already discussed. As we will see this protein reappears a little later; its name is lysyl oxidase-like 1 (LOXL1).

Studies have shown the importance of this cross-linking in the perineal organs. The uterus, the colon or the urethra are also potentially impacted. But to return to the perineal space, this experiment in mice with a defective elastic capital due to a cross-linking defect illustrates the inconveniences that many people experience due to aging, such as vaginal leaks or bladder insufficiency, for example. One can only agree with the physical strengthening exercises that are recommended to fight against the weakening of the tissues that accompany these organs. In addition, to remedy vaginal prolapses, sometimes synthetic membranes are used to support the walls of the organs. Although generally considered biocompatible, the use of these tissues can cause problems in the long term because of their less elastic characteristics than the original tissues. There are risks of fibrosis around certain implants. We find the need to develop truly biomimetic elastic tissues, i.e., mechanically similar to the treated tissues. These developments are very important for women, and even more so for those who have a mutation affecting their fibers, as in the case of cutis laxa when childbirth is on the agenda.

2.8.2. *Mechanical adjustment to pregnancy*

The development of a child's body in another body is not trivial and requires considerable adjustments. Among the many hormones that control this development, relaxin is involved in the biomechanical adaptation of the mother. This hormone belongs to the family of insulin and insulin-like factor (IGF1) family mentioned above. Its action, supported by other hormones, results in a slightly more flexible and lax body. Although sometimes painful, the induced increase in joint laxity and intervertebral discs is intended to facilitate pregnancy and childbirth. For example, a Brazilian colleague of mine from the University of Campinas observed a unique formation of elastic fibers on the symphysis pubis ligament a few days before the delivery, which enabled facilitation (Castelucci et al. 2018). Relaxin appeared to be at work in this experiment. This observation made in mice could not be corroborated in practice in women, as we agree that the moment is not very conducive to this type of analysis. This biomechanical aspect of childbirth completes the evolution of the cervix and its programmed opening by the action of the uterine muscles. The cervix itself is poor in muscle fibers. On the other hand, it is rich in connective tissue and fibers. It maintains this tonic structure throughout the pregnancy and undergoes modifications during labor. There is a decrease in collagen fibers and an increase in connective tissue that promotes hydration. The result is a softening that allows the cervix to dilate, leaving only a thin muscular diaphragm open to the vagina which in turn will have to undergo a tremendous extension. The practice of elastic training of the tissues, muscles and ligaments of the perineum is sometimes proposed in obstetrics to improve the extensibility of the vaginal walls.

With regard to pregnancy, one of the fears is that the extension of the belly's skin is accompanied by stretch marks. These correspond to an unsuccessful attempt by the dermis to adapt to a significant extension under the constraint of hormonal deregulation. The cells of the dermis then begin to produce elastic fibers but as they no longer know how to do this well, the result is an amorphous mass of poorly tied fibers that give the appearance of striations demarcating the lines of extension. Several solutions have been proposed, but few have been fully validated to avoid these marks, other than to perform directed massage and pressure to reduce the internal tensions of the skin (Korgavkar and Wang 2015).

Much remains to be studied on the mechanics of sexual activity where the amount of biomechanics research at the macromolecular scale has been relatively little documented. Sometimes, COVID-19 has yielded surprising information that helps research. Among the range of unusual actions of the virus is an increased risk of male infertility for infected individuals (Younis et al. 2020). The reason for this is the very strong presence of the ACE2 protein, which is the virus' entry point, in the cells that support spermatozoa formation. These cells are important in the formation of male gametes, and their deregulation and dysregulation are often synonymous with infertility. Apparently COVID amplifies the risk. This convergence will remind some of the infertility risks associated with testicular inflammation in males who contract mumps (caused by another RNA virus). It also introduces the mechanics of sperm formation and release which presumably depends on tissue structure. It could thus be considered that the decrease in male fertility could be linked, in one way or another, to the decrease in the elasticity of the walls of the different ducts (epididymis, vas deferens, urethra) or the weak swelling of the spongy bodies of the penis by vasodilation. Male virility needs its elastic capital.

2.9. The microbiota and its body bioreactor

In our body envelope, we cohabit with several million inhabitants, all of which have their own specific requirements and which can be beneficial, but sometimes toxic. The quality of the microbiota has indeed become an essential parameter of health, for better or worse – for the better by complementing our metabolism (Lazar et al. 2019) and for the worse during parasitic or infectious infections, particularly in sexual transmissions or following surgery or prosthesis. One of the major issues in the fight against these infections, in addition to drug resistance, is the possibility presented by some species to develop biofilms at the infected sites, which will make them more difficult to access to body or chemical defenses. This is for example the case for the fungus *Candida albicans*, which is very present in the population and is quick to cause candidiasis in case of immunodepression.

We are therefore a kind of bioreactor for our host microorganisms; we provide them with heat, liquids and nutrients. They benefit from our internal environment whose parameters are updated at all times in humidity, salinity, acidity (pH), temperature and pressure, among others. Let us take for example the two main classes of bacteria: those that need oxygen (aerobic bacteria) and those that do not (anaerobic bacteria). They do not proliferate at the same time and generally not in the same places, unless the environmental conditions change. Without digging very deep into the body, we can take the example of bacteria in dental plaque. Aerobic species proliferate after brushing when plaque is removed. If the plaque thickens, due to negligence of hygiene or due to the production of materials by sugar-fed microbes, the lower, non-oxygenated layers of the plaque will see the prevalence of anaerobic bacteria likely to attack the gums. They will produce acid which will lead to the dissolution of the calcic materials of the enamel which is the strongest material in the body. We accelerate the work of these decalcifying bacteria by frequently drinking the very acidic and very sweet sodas that are part of the worldwide pandemic of cavities. The love of certain bacteria for our overconsumption of sugars is limitless!

Both classes of bacteria are therefore found but not necessarily in concomitance. The growth of aerobic bacterial colonies is promoted by an agitation that brings oxygen and nutrients. We have seen previously how much our colonies are subject to pressure and propagation waves. The intestines are very rich in bacteria. These help digestion in the small intestine and then fermentation in the colon. Timing is everything. Some bacteria may be at cross-purposes, such as colon bacteria that colonize the small intestine. If they swarm there, they can cause an early fermentation of food, which disadvantages their assimilation. One swells without benefiting. This imbalance is associated with small intestinal bacterial overgrowth (SIBO) syndrome, which causes bloating, inefficiency of assimilation and digestive discomfort. There is no doubt that the laziness of the peristaltic movements of the intestines can facilitate this type of synchronization error. From all this emerges the concept of a dynamic interaction of our microbiota with our internal mechanics, conditioning the balance of the populations of our microbial passengers that are anything but clandestine. As in a bioreactor, the better the agitation allowed by the elasticity of the systems, the better the conditions for beneficial bacteria.

Some of our personal bacteria have adapted well to us because they have patterns on their walls that resemble or promote association with our tissue proteins. This is the case for streptococci and staphylococci, which have become experts in camouflage by appropriating the patterns of host tissue, i.e., us. They can attach themselves to heart and bone tissue, and even induce an autoimmune response. This is for example the practical implementation of "the badly-treated-cavity-which-reaches-the-heart" when oral streptococci squat by biomimicry on the walls of the heart and generate an infective endocarditis (Sommer et al. 1992).

If our mirrors were equipped with a high-definition microscope, we would see that our entire body envelope is full of bacteria and fungi. Hundreds of species make themselves at home by choosing their favorite places. They select their temperature, the level of humidity, the local acidity or sweetness, as well as the richness of the fragrances exhaled by our secretions. The more varied the flora, the more they protect us from external and internal aggressions by collaborating with the immune system. Sometimes, they exaggerate and it will be necessary to limit their growth while checking that their playground (i.e. us!) has nothing to reproach itself. We understand that we have to cajole them. For example, for the skin (which is a carrier of peaceful bacteria), we must use washing products that preserve the physiological parameters of the skin (gentle pH, little aggressive soap, moisturizing) and non-toxic cosmetic protections for our hosts. On the other hand, we need to find ways of dealing with a multitude of aggressive microorganisms, such as this black fungus that induces the death of the tissues it contaminates, which gives the black color of dead tissues (we speak of necrosis) associated with the opportunistic mucormycosis that takes advantage of the weakened condition of those affected by COVID-19 (Skiada et al. 2018).

The SARS-CoV-2 pandemic has also focused on the microbiota of the lungs and intestines. It was already known that microbiota imbalances are often associated with acute phases of pulmonary failure, requiring the help of an artificial respirator. Among these imbalances, the lack of certain bacteria modifies a major brick in the environment of the intestinal and pulmonary walls. This is heparan sulfate, a molecule related to heparin, which is used by the virus to infect cells in the body (Clausen et al. 2020). As a result, the absence of these bacteria promotes infection. So let us take our microbiota into account through proper management of our elastic bioreactor that will provide the proper mechanical growth parameters for the microbes we know are beneficial.

2.10. Conclusion

We have seen through these examples how our life takes place in a liquid environment, where everything communicates with everything. In the 17th century, with the discovery of blood circulation, the body could be considered a great distillery, which was opposed to the man-machine vision. We now know that both solid and liquid aspects coexist and collaborate. We will add to this the psychological aspect that we will see later. While waiting to discuss the interaction of matter and spirit, let us be prosaic and attach importance to the color, density and odor of our internal productions, at the same time reflecting our mechanics (material and fluidic), our energetics and perhaps even our psyche. Uroscopy, the science of urine analysis, which had great success in the previous centuries, has fortunately been replaced by medical analysis. It is up to each person to find the 20 shades of yellow described in the 13th century or to look for a sweet fragrance in these productions to suspect diabetes. All this is less precise than a laboratory examination, but in the end, it completes the information shared by the mirror and the tension points of our body.

To sum up this chapter, we can now better understand our relationship to fluids. They circulate continuously and diffuse between cells, vessels, nerves, tissues and muscles and even bones. We must get away from the idea of perfectly sealed vessels and structures. On the contrary, everything communicates through pipes and walls that are more or less porous depending on the location and pressure. Not enough water and the system slows down or even dies. Too much water and the system suffers and drowns. Water is essential to life in general and to elasticity in particular. Without water, the elastic fibers lose their power and nothing will work anymore. It is enough to have in mind the image of an old cracked rubber band recovered at the bottom of an old drawer to make a rather correct representation. It is the same thing in the body – in the very short term for the lungs, the heart, the kidneys and the liver, and in the longer term for the cartilage, the skin, the ligaments, the fascias and the bones. So we drink, regularly and reasonably, and we limit alcohol and sodas which damage the kidneys, the liver and the blood vessels. They also slowly but surely afflict all our senses. This is what we will see in the next chapter.

3

Elasticity and the Senses

Figure 3.1. *The sensory bullfight. This fourth painting illustrates the abundance or lack of sensory perception. I chose to illustrate this sensory merry-go-round on the basis of* Study for a Bullfight, No. 2 *by Francis Bacon by also introducing a variation on* The Blind Leading the Blind *painted by Pieter Bruegel the Elder. The giraffe represents an astonishing evolution of this sensoriality which deserved a place in this circus (Gurevich and Gurevich 2015)*

3.1. Introduction

When we talk about mechanics, we are talking about measuring and controlling the movements made possible by mechanics. The immediacy of

---

For a color version of all the figures in this chapter, see www.iste.co.uk/sommer/elasticity.zip.

the response is often required, especially when a danger must be identified and transcribed into effective actions.

We instinctively know several ways to estimate the elasticity of our body. Vision and olfaction give us information about the thickness and tension of the skin as well as the colors and odors of urine and feces. Touch, through palpation, allows us to feel the elasticity of the skin and subcutaneous tissue. In a less obvious but daily way, we find the mechanisms of palpation when chewing because ligaments around the teeth are provided with biomechanical sensors. Of course, modern quantification technologies are more precise, but the sensory self-analysis that everyone does is important. This is also what people who have lost all or part of the use of a mechanical function learn, with the pain often associated with it. Sensations are sometimes difficult to identify and translate, as I have found with colleagues sharing a CNRS interdisciplinary challenge on sensory deficit as well as at several European networks on aging that I have led.

There is a reciprocal link between the senses and elasticity. It is well known that the senses serve to control and measure movements and actions, which depend in part on mechanics and thus on elasticity. But it is much less known how much the senses themselves depend on this elasticity. This is what we will present in this book, starting with singing and dancing, which mobilize both aspects of this relationship. Then, we will look at each of the senses, some of which have been highlighted by COVID-19 as being altered.

3.2. Singing and dancing

To understand the idea of a link between elasticity and sensoriality, there is nothing better than to consider the mechanisms of singing. So we should sing, frozen with fear or not, whispering or at full voice. And if possible, we should do it in front of a mirror that will bring us additional and unexpected visual information!

Singing, in terms of physics and mechanics, produces a displacement of the air that propagates like a sound wave, like the waves that ripple on the water after the impact of a stone. At the vocal level, the impact of the stone in the water corresponds to the movement of the vocal cords which strike the air, more or less quickly according to the height of the sound. The vocal cords are folds of laryngeal tissue consisting of a majority of collagenous tissue at the base and a richer mixture of elastic fibers and a type of loosely cross-linked collagen at the surface (Moore and Thibeault 2012). The alteration of the fibers over time explains the drop in pitch toward the lower tones of the voice as the vocal cords become less mobile and vibrate less quickly. Heavy smokers, with degraded vocal cords, no longer offer anything but the hoarse harmonies of someone used to propping up a bar every night, which is something that struck me when I heard the girl with the big black eyes at just 10 years old.

To sing, you need to propel air efficiently, which brings us back to the benefits of belly breathing that we have already mentioned. We mobilize the diaphragm, the abdominal and thoracic muscles, the fascias, the vocal cords (we speak in fact of vocal folds), the epiglottis, the tongue and the twists and turns of the nasal cavities. Is that all? Oh no! There are the feet that ensure a good posture and the perineum which provides support. And when the air tickles the pharynx, the larynx and the epiglottis, it activates the vagus nerve which sends soothing messages from the sympathetic system. The ears control the tone and incite the brain to adjust the tension of the vocal cords through the laryngeal cartilages. Singing in tune is therefore a simple matter of adjusting the tension. So, there is no excuse not to sing and to mobilize almost all the elastic properties of the body, in one go and at little cost. You can sing alone or sing in a group, which also activates oxytocin, the social hormone.

Sociability is also drawn upon in the practice of dance. I had the chance to participate in self-evaluated dance workshops as part of this CNRS challenge on perceptual disability. The Labodanse laboratory (CNRS – Université Paris 8; Joufflineau et al. 2018) that had developed tools of symbolic and interactive projection of dance had invited us with their colleagues from the COMETE laboratory (INSERM, Université de Caen). It was a contact dance between people suffering from a hypermobile type of Ehlers-Danlos syndrome whose precise genetic cause remains to be identified. For these people, their hyperlaxity is such that it disturbs their posture, movements and positioning in space. They then struggle to give the appearance of a completely normal life and frequently resort to support garments to sustain their position. The experience was extraordinary but difficult for them, so much so that they had to compensate for a structural deficit by mobilizing all their senses in order to manage the situation as best they could. So let us see how the six senses work together with the elastic properties by firstly, to help the reader, presenting a small diagram showing the main structures used by three of these senses.

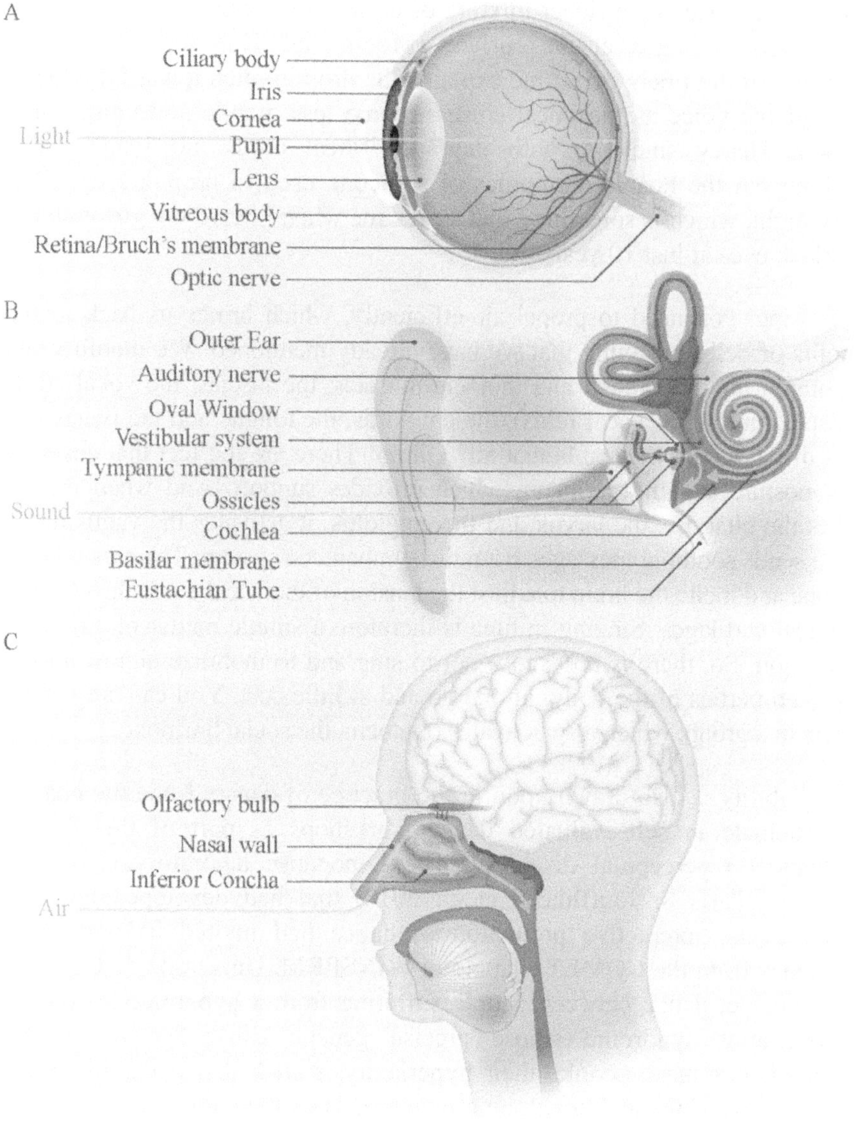

Figure 3.2. *The elasticity that is essential to the hearing, vision and olfaction pathways*

3.3. Light transmission and elasticity

The photons of light pass through the cornea, the aqueous humor and then the pupil and the lens. They pass the vitreous humor of the eye cavity and impinge on the cells of the retina. The energy of the photons is transformed by the cells of the retina into an electrical signal that is conducted to the brain by the optic nerve (Figure 3.2).

The light therefore diffuses to the retina where it is translated into an electrical impulse. Obviously, the cornea, the lens and the aqueous humor must be as transparent as possible. The cornea is a thin, almost transparent membrane. As for the lens, which enables the focusing of light on the retina, it must be very transparent even though it is one of the densest structures in the body. The specific arrangement of collagen fibers that constitute it behaves like a kind of liquid crystal, as demonstrated by colleagues from Lyon's Institut de Biologie et Chimie des Protéines (Torbet et al. 2007). The plasticity of the lens, its maintenance by elastic fibers (the ciliated body) and muscles allow the light path to be focused. The partial opacification of the lens, synonymous with cataracts, is still the first cause of blindness in the world. The glycation of proteins and collagen is one of the major causes of this opacification (Rowan et al. 2018). The harms of excess sugar are limitless.

Blindness is also known to be induced by alterations of the retina or of an underlying membrane called Bruch's membrane. The latter is very rich in elastic fibers and therefore subject to glycation, which was one of the causes of blindness in untreated diabetes. The retina is also covered with blood vessels whose alteration contributes to macular degeneration or glaucoma. In the latter case, a biological marker has been identified, whose loss of functionality has been associated with a susceptibility to an increased risk of glaucoma. In the case that was the subject of a collaboration with colleagues from the University of Cologne in Germany, it appeared that one of the enzymes that ensures the cross-linking of elastin could be deficient (this is again LOXL1, which has been found before). This results, among other things, in vascular problems in the retina (Schlötzer-Schrehardt et al. 2008).

In the field of perception, vision pays a high price to modernity. The tools of visual focusing and adaptation become less efficient. Old histological observations described a strong elastic component around the optic nerve, at the site of its connection to the eye, that allows very fluid rotations of the eye. We have mentioned the elasticity of the lens, as well as the elastic filaments and muscles that position it and allow adjustment of focus. The eye is thus subjected to incessant movements, some of which are at the limit of what can be detected, such as the rapid saccades that allow us to pursue an object. The sedentary nature of the eye and focusing on a screen are thus very much involved in the pandemic of myopia and presbyopia. The preventive solution is simple: get some fresh air and confront your vision with the great three-dimensional horizon of nature!

3.4. Auditory transmission and elasticity

The perception of sound deserves to be examined from a mechanical point of view. Hearing is first of all the transmission of a vibration of the air that must be channeled, transmitted, amplified and then translated into electrical impulses for our brain. The vibration of the air is channeled by the external ear, transmitted to the eardrum, amplified by the three ossicles in the internal ear which reflect their vibration to a kind of second eardrum, the elastic oval window. It is thus in reality a sum of vibrations which come to break on elastic membranes. The tympanic membrane is made up of collagen fibers very enriched in elastic fibers in their center, a little like certain tambourines composed of two types of concentric membranes. The eardrum vibrates and activates a chain of ossicles which knock on the door of another membrane, which in turn transmits its vibration to the liquid of the cochlea whose movement will excite the nerve cells which will communicate their influx to the auditory nerve. The vibrations propagate in the inner ear and induce an electrical response from the cells in the walls of the cochlea, while an underlying elastic basilar membrane undulates to amplify all this. The liquid movement will continue to die out upon contact with another elastic membrane, the round window, at the exit of the cochlea. In total, the initial sound is transmitted to the brain with an amplification of about 20 times.

The eustachian tube balances the pressures in the ear with atmospheric pressure, which we also do by blowing and pinching our nose to balance the pressures on either side of the eardrum. The position of the vestibular domain (used for balance) is shown in Figure 3.2.

All this is excellent, but fragile. Loss of hearing often concerns the capacity to produce the electrical impulse, which leads to the need for an apparatus to transform the sound into direct electrical stimulation. Deafness can also come from the alteration of an element of the mechanical chain, elastic membranes and small bones which might need to be replaced by biomaterials, according to the circumstances. This is a clinical sign for some people with syndromes such as cutis laxa. It is also a sign of our civilization, that we have been exposed to sound sensations that are much too loud, either by obligation or by choice. We are often far from having the hearing of hunter-gatherers. But in any case, one must continue to practice, if possible, in order to maintain the mechanical part of the hearing as long as possible. The elasticity of hearing is like that of the rest of the body; it only deteriorates if you do not use it or use it badly!

3.5. Olfaction

It was not obvious to integrate olfaction in this elasticity affair and in fact it is the loss of this sense in COVID-19 that brings me there, since in a roundabout way it is true. It has been a while since SARS-CoV-2 fully entered this trial. Its effect on the lungs is significant. But its first point of entry can be at the nasal level.

Air is inhaled through the nasal passages, at the level of the nostril. It then infiltrates the various nasal cavities to reach the trachea and bronchi. The virus brought by the air can enter the nasal mucous membranes through the preferred entry point, the now famous protein ACE2. But if it has a nose for it, the virus can continue its way and penetrate to the olfactory bulb that is located at the bottom of the nasal cavities, almost under the eyes. This olfactory bulb consists of a set of sensory cells that will detect the chemical composition of elements that pass within range of the nose. To increase one's sense of smell, one must direct and retain the flow of air toward this sensory organ. The particularity of the olfactory bulb is that it is directly connected to the brain, which is facilitated by its positioning just below.

There is an amazing mechanism in the nose that is used to humidify and eventually warm the air to protect the lungs. To do this, the nasal mucous membranes are humidified because of the alternating turgidity of the mucous membranes at the base of the nose tip. In a cycle of about 3 h, blood vessels dilate and induce humidification and swelling of a nostril's tissues. At the same time, the vessels of the other nostril are tightened by vasoconstriction and the tissues release their humidity, which allows the optimal opening of this nostril and the humidification of the inspired air. This cycle is observed continuously, under the control of the peripheral nervous system, such as the vagus nerve for relaxation and the sympathetic system for constriction. Some yoga exercises take advantage of this mechanism of alternating vasodilation and vasoconstriction of the nasal mucous membranes. The solicitation of the vessels of these mucous membranes makes it possible to induce the formation of nitric oxide, which is a molecule that will intervene everywhere in the body to help transfer oxygen and carbon dioxide and promote energy production. Nitric oxide is also an excellent physiological cleanser for our nostrils, which we will help to diffuse in the sinuses by nasalizing a sound like the famous "om" of yoga. So, we breathe as much as possible through our nose, especially when we are sure that no unfortunate virus is lying around!

3.6. Taste

We have seen three of our senses that are duly concerned with elasticity. This is not obvious for taste which does not seem to have been studied this way. It is however on the agenda because ageusia is also a manifestation of COVID-19. We know that there is elasticity at several levels in the mouth. There are elastic fibers and collagen fibers in the lips. This must undoubtedly annoy plastic surgeons who only have collagen or hyaluronic acid to fill lips; they cannot add elastic fibers which alone would ensure the flexibility of the mouth without risking the improbable duck lips. These elastic fibers can also be found on the tongue or as components of different cartilages, without being directly linked to taste. Dedicated research would surely complete the picture because oral mucosa, mandibles, gums, teeth and their periodontal ligament are all examples of mechanical perfection.

3.7. Touch and proprioception

The impressive dashboards of airplanes and other sophisticated machines are simple compared to the control tower of our body. Let us think about a control network with sensors at every level that are linked to the brain, to associated structures such as the enteric brain in the gastrointestinal system and to muscle motors. The skin is thus full of tactile receptors reporting on local pressure, texture, shape, the grasping of objects and temperature. In addition to the injunctions of movement and grasping, the different nervous circuits and their cutaneous terminals report on heat or pain in order to complete the data coming from touch. The mechanical environment is omnipresent since nerve endings are associated with elastic tissues that age (Decorps et al. 2014). The identification of nerve relays between these receptors and the sensorimotor areas of the brain were the subject of the Nobel Prize in Medicine in 2021 awarded to David Julius and Ardem Patapoutian for their discoveries of temperature and touch receptors.

Elasticity is also involved in proprioception. This sixth sense is a fundamental factor of perception to avoid falls, injuries and optimize movements. Proprioception is fed by sensory sensors positioned under the skin surfaces of the body as well as in the muscles, ligaments, tendons and fascias. It participates in this flow of information and instructions that allows the body to situate itself at any moment in time and space. Integrating also the information of the vestibular gyroscope of the cochlea, proprioception allows us to live in our gravitational world.

Post-traumatic and infectious alterations can have a negative effect on touch and proprioception. For example, at the Institut des Sciences du Mouvement in Marseille, colleagues are following the extremely rare case of a woman who has lost almost all sensory connections between her head and the rest of her body due to an infection in her youth. As a result, she no longer has any proprioceptive or cutaneous sensation of her body below the neck, which she replaces with visual control (Miall et al. 2019). As a result, in the dark, she does not know where her legs, arms, and body are. As she says, in the dark, "I'm just a brain in the air"! It's impressive to see her raise her glass with serenity and control, knowing that only the shape of her fingers crushing on the glass lets her know she's holding it. If you ever see someone drinking their coffee out of a wine glass, perhaps it does not reflect snobbery but a visual and haptic necessity.

There is a relatively frequent situation where we can find ourselves in this case. These are circumstances where our vestibular control is compromised, when we are land or sea sick because the vestibular fluid is shaken in all directions. The crystals that it contains and that serve as gyroscopes can no longer align with the corresponding proprioceptive information of the body, and we are left with only vision to position ourselves. It is better to look far ahead at a fixed point when you are seasick.

There is an entire panoply of receptors regulating the autonomous movements of the body. There are receptors for mechanical pressure and those for chemical pressure. In this way, information is continuously sent to the brain to control breathing, heart rate, renal control or stature, as well as the level of carbon dioxide in the blood, the measurement of blood pressure, vascular tone or the verticality of the body. All this leads us to take a few steps towards the links between the peripheral nervous system and elasticity, often at the limit of current studies on the transmission of nerve impulses.

3.8. Elasticity and the peripheral nervous system

While the body and the mind are intimately connected, this is essentially because of the diffusion of nerve impulses between the central and peripheral nervous systems. The brain is the central nervous system. Partly, the peripheral nervous system begins at the spinal cord and also involves all the nerves that run through the body connecting to the brain. Many people undergoing anti-tumor treatment have a painfully intimate knowledge of this peripheral nervous network, which always comes with associated sheathing and logistics. I learned from my histological colleagues at the Institut Pasteur in Lyon that nerves are accompanied by blood vessels and lymphatic channels. All these canal structures are themselves accompanied by a continuum of solid and elastic tissue bathed in a constant flow of liquid.

When a physical signal such as a caress reaches a nerve cell, a wave of ion displacement moves along the membrane of that neuron and spreads to other neurons. The flow of ions "skips" the areas that are surrounded by a myelin sheath. This speeds up nerve transmission. Like other cells, neurons are supplied with fluid, salts and nutrients by the surrounding blood and lymphatic vessels, while their cellular waste products are carried away by lymphatic channels.

The complex composed of nerves accompanied by its perineurium (vessels, lymph, accompanying tissues) can be compressed by mechanical pressure. In addition to pain, this compression can also have a negative effect on the logistics of these nerves and therefore on the transmission of their signal. This is for example the case of the nerves that emerge from the spinal cord through the spaces between the intervertebral discs. The nuclei of these intervertebral discs are made of fibrocartilage rich in collagen for one part and elastic fibers for another part. This means that they are sensitive to compression, inflammation and normal and accelerated aging (for example, by an excess of sugar or oxidants). This is the case for joint pathologies when the nerve pathways are compressed by repetitive movements, such as carpal tunnel syndrome in the hand and Morton's syndrome in the foot.

Although little studied as a system, research on the perineurium is seeing its importance grow in neural reconstructions (Kucenas 2015). Alteration of this sheathing and tissue accompaniment around nerve fibers could be linked to many pains and dysfunctions. Myelin is surely the best known of the nerve sheathing structures, as its deficiency is linked to multiple sclerosis. But other more diffuse pains are described by people suffering from genetic diseases of elastic fibers, collagen or other matrix proteins, as well as the whole spectrum of clinically anonymous pains that refer to terms like fibromyalgia or algoneurodystrophy (Sprott et al. 1997). Gentle movement practices can therefore be considered beneficial when pharmacology reaches its limits, to restore tone to this perinuclear system, taking care not to worsen the pain. We should make our nerves do soft gymnastics, as we make our blood vessels, our muscles and our joints do Pilates!

The example of the perineurium punctuates the first chapter on the omnipresence of elasticity in the body. The exploration of these elastic functionalities could go on for a long time, but I believe (and hope) that we have understood the message: the body needs elasticity at all levels and few functions escape this mechanical necessity. Interdisciplinarity is required as well as the scales to be considered, at the level of the skeleton, the organs, the covering tissues (skin and cartilage) and the fascias running through the body as well as the transmission of mechanical information to the nuclei of the cells. The structures ensuring optimal elasticity are hardly synthesized after growth, leading to regressive or even disabling situations to which the body must adapt. It is now time to present the challenges that concern the decrease or weakness of elastic capital.

PART 2

The Four Challenges of the Elastic Human

Writing an essay on the functioning of the body always has the intention of curing or paving the way for good practices that do good or prevent harm. In reality, the needs are innumerable; so many bodies have suffered, are suffering or will suffer from dysfunction linked to an elastic capital defect, inborn or acquired! Our purpose is to share some beneficial actions for our elastic capital by supplying information that depends on specialists but also on our own behaviors. I will only focus on the areas that concern the elastic system, and not being a healthcare worker (although I have worked a lot with them), I will always refer to their fields of competence.

Depending on the case, the disease imposes its urgency, its potential chronicity or the need for prevention. The time of the story is not then on the same scale. Cardiac arrest, stroke, pulmonary fibrosis under oxygenation, embolism, phlebitis, major burns or major bone fractures impose an emergency. Chronic diseases and non-lethal genetic defects take a longer period of time. Rheumatism, ankylosing spondylitis, osteoarthritis, organ fibrosis, emphysema, vascular tortuosity, heart murmur, intestinal sluggishness and diverticula, kidney stones, sensory deficits or Parkinson's disease, among others, imprint their painful chronicles over the very long term and can truly spoil life.

Research for understanding and improving treatments is inherently multi- and interdisciplinary. Medicine, pharmacy and engineering for health are complemented by other care, in the literal sense of caring for and looking after the well-being of people with stress, which we will see in Chapter 4. In Chapter 5, the balance between anabolism and catabolism will be discussed. We will talk about nutrition and pollution, vast subjects that concern everyone. Movement is the key word for maintaining one's elasticity or limiting its decrease. This work of behavioral prevention orchestrated by movement will be discussed in Chapter 6 to delay the moment of becoming immobile for eternity.

Finally, in Chapter 7, we will explore the links between the laws of elasticity of the body and the mind. And there is much to consider, since the defects of elastic capital lead to a decrease in our "capabilities". This will be the domain of the individual mind, but also of social and societal regulations that allow us to remedy the defects of our capabilities, if not to regenerate them. The ambition of the fourth challenge of the elastic human goes far beyond the limits of the body to address ethics.

4

The First Challenge for the Elastic Human: Mechanical Stress Management

Figure 4.1. *Repairing the elastic human. The fifth painting integrates the history of engineering for health between surgery, pharmacology and therapeutic follow-up. The figure is based on the painting* The Insertion of a Tube *by Georges Chicotot, painter and head of the radiology laboratory in 1914 at the Hérold Hospital (Paris). A nod is given by introducing a character who does not always show all his cards (*The Cheater with The Ace of Diamonds *by Georges de la Tour), because the elastic human is well aware of a physiological and mechanical complexity of which we do not know all the cards.*

For a color version of all the figures in this chapter, see www.iste.co.uk/sommer/elasticity.zip.

4.1. Introduction

We will therefore talk about stress management in this chapter. We will restrict ourselves at this stage to physical stress, which applies to solid structures, disturbs the fluids and gases of the body, but also is determined by our innate and acquired capacities. We will not forget the psychic stress, of course, but that will be covered in the last chapter.

Research is confronted with a major problem, that of the heterogeneity of the stresses and the scale to be taken into consideration. It must be admitted that the complexity of the living world induces a slight fluctuation between the different approaches. "Imagers" and biomechanists focus mainly on tissues and organs; biochemists and biophysicists explore molecular assemblies and interactions; biologists reframe divergent or ineffective cells, tissues and organs; physiologists and geneticists deconstruct their relevant models; clinicians and pharmacists deal with pain, inflammation, infection and trauma; electronics engineers and physicists develop implantable or accompanying devices; and sociologists, psychologists and philosophers combine all these personal and social behavioral data to qualify behaviors and their deviation from the dominant stereotypes. We must therefore deal with this complexity, often at the limit of our knowledge, which feeds a daily doubt, a doubt that the general public discovered with the constant minute-by-minute monitoring of a global pandemic whose truths are permanently being constructed.

4.2. Stress of elastic structures

There are many types of pain resulting from the failure of elastic capital, and they are mentioned in many works of world literature and filmography. Among these, the testimonies of life with war maiming are in first place. The text of a song sung by a former French soldier of the 1914–1918 war sums up perfectly this difficulty of being "ill". This French soldier was named Gaston Ouvrard, but it could be any soldier wounded during the conflicts that are still going on around the world or any disabled person, whatever the cause. But his song is worth analyzing, first because he was singing with knowledge of the facts, as he had been doubly wounded, and second because of the evocation of traumas, most of which fully concern the elastic system. His text begins (translated here) with the ills that concern the regulation of fluid mechanics, and thus the organs and soft structures such as "the spleen that dilates, the liver that is not straight, the pylorus that is colored, the stomach that is too low". Then, he insists on the solid structures in particular: "the ribs which are much too high, the hips which are dislocated, the thorax which is out of alignment, the knees which are soft, the ankles which are twisted, the kneecaps which undulate, the toes which are not the same".

Let us start by looking at how modern-day medicine knows how to diagnose the "dilating spleen". The identification of this disease is currently facilitated by internal body visualization tools. We can only welcome the precise visualization of splenomegaly when I have an "enlarging spleen", because of the improvement in the consideration of the elasticity of materials and the mathematical processing of signals, notably by ultrasound and magnetic resonance imaging (Gennisson et al. 2013). Diagnosis can also be completed using various imaging modalities (magnetic resonance imaging, computed tomography, ultrasound, radioactive or chemical marker tracing) that enable discernment of the topology, composition and mechanical characteristics of soft, elastic and non-calcified tissues, such as skin, ligaments, fascia, organs, brain and sensory tissue. This diagnosis completes the visualization of calcified tissues that has been known for a long time because of X-rays. We can then really evoke the notion of a "transparent human". Several generations of grandparents can attest to the spectacular progress in the definition of in utero ultrasounds.

It is now understood that connective tissue is sensitive to stress. It is even a key element of aging at all levels, in the cells, the tissues, the body and with significant repercussions on the senses and the state of mind. Sources of mechanical stress are abundant, but engineering for health, which includes imaging, has greatly improved the management of disease. These advances make it possible to detect the presence of fibrosis in COVID-19 patients. This is not to be congratulated for the sake of medical achievement, but it must be admitted that humanity was very lucky, in its great misfortune, to experience the SARS-CoV-2 pandemic in 2020, when many of the tools that were used to identify and contain the pandemic were able to be deployed. The situation would have been much worse a decade earlier.

These advances, and in particular the obtaining of three-dimensional images, have been accompanied by the development of prostheses by computer-guided additive synthesis of three-dimensional objects. The widespread use of 3D printing is used to build a chair, car part, tooth or bone. In addition, the development of biomimetic biomaterials, inspired by what nature has developed. These biomaterials serve as a niche for stem cells for a guided reconstruction of biological tissues. We can speak at this stage of body hybridization in order to restore a certain functional capability. It is a concrete and beneficial application of the "repaired human" concept when it comes to compensating for deficient human capacities. The ethical questions about human augmentation are of a different order, as it seeks to increase our current potential towards an "augmented human", to go beyond a state provided by thousands of years of evolution. In the current state, this desire of augmentation seems to me less like an exercise of public health than a desire to appear, like a form of hypertelia, when structures and organs take on disproportionate proportions (Tort et al. 2020). It is a bit like when a deer with long horns or a peacock with a long tail strut to denote concupiscence. We return to the discussion between the ingenious Daedalus, who created wax wings to escape the labyrinth, and his son Icarus, whose mad desire led him toward the sun where he burned his wings.

One of the technical limitations of the formation of prostheses and orthoses lies directly in the notion of elasticity. We now know how to make many hard materials, such as a hip prosthesis or a tooth. We must also recognize and encourage the enormous progress in the development of metal and calcium prostheses that replace bones and even joints such as the knee. Gone are the stiff prostheses that creaked with every step! However, there is still a lot to be developed as soon as we approach something softer and more elastic, but it is certain that one day we will surely replace the menisci, the tendons or ligaments by biomimetic biomaterials equivalent to the original tissues. These biomaterials will also be used to reproduce identically an intervertebral disc with its external fibrous ring and its internal pulpy nucleus, gelatinous and somewhat elastic. However, there is an urgent need for this because millions of damaged spines are waiting for a new lease of life!

Since the body loses its elastic capital, the simplest thing would seem to be to replenish it. This is the first question asked by the families of children with elasticity defects. It would then be a matter of providing materials that have the same elastic characteristics as the original tissues. However, it is impossible to extract elastic fibers from tissues for medical applications. The development of elastic biomimetic biomaterials is obviously being studied all over the world, and in particular by my former teammates in Lyon who are continuing my work (Carrancá et al. 2021). This would have multiple applications to manage vascular and pulmonary problems, bone repair and skin, or the reconstruction of vocal cords, to name just a few examples other than the appropriate reshaping of our wrinkles.

Research is progressing to provide the body with sufficient elastic capital or to encourage it to do so at the right place and at the right time; this is exactly what is needed in burn victims or at least in reconstructive surgery. The use of adult stem cells is already omnipresent, whether in clinics or in cosmetic surgery. It is a matter of using the cells that serve the normal renewal of the skin and reinjecting them in the places where it is necessary, respecting the mechanical tension lines. We classically use fat cells from the skin of the belly to extract stem cells. And if there is one tissue that is not in short supply at the moment, it is fat!

In the meantime, existing techniques are already very useful for repositioning a vertebra and attempting to re-innervate the limbs of a paraplegic, as well as for covering large burns with artificial skin. Fortunately, the majority of people do not have recourse to these sophisticated technologies and the surgical introduction of hip or knee prostheses have become routine applications. However, there are still many challenges to technically manage these stresses, such as less aggressive knee replacement or challenges regarding artificial organs, inserted or not in the body. In most cases, the notion of elastic capital can be invoked, obviously in addition to other concepts and techniques. The topicality of this issue is once again evident, since an urgent challenge is to reshape the walls of the pulmonary alveoli deteriorated by fibrosis.

4.3. Stresses on organs and fluids

The ultimate toxic behavior for elastic capital is poor water management. As we are over 60 percent water, this component and any variation in its liquid volume is significant. This is especially true for our fibers, which need to be directly bound to water to be functional. Water is the resource of life; we should never forget it, since it is one of our assets if not the most precious one that we waste and cheerfully pollute nevertheless. The body is made of about 40 L for a "not too dry" individual of 70 kg. More than half of the water is cellular and necessary for the metabolism. Nearly a quarter is found in the interstitial fluid, which demonstrates its importance, while the rest is divided between the blood and lymph. We also find fluids in the digestive system, the eyes, nose, ears, cerebral spinal fluid and our various secretions. We have seen that water diseases are often associated with stress of the kidneys and bladder, when they are supplied with too much or too little liquid, or when this liquid is too dense, rich or toxic. In fact, all causes that clog, leak, hold back or cause the cardiovascular, lymphatic and intestinal flows to flow backwards are causes to be considered and remedied. Medicine manages these stresses in emergency by renal dialysis, cardiac rhythm regulation with a pacemaker or the placement of stents and other catheters (Peters et al. 2020). But the first step, which depends only on each individual, is to consume about 2.5 L of water per day to satisfy the proper water balance between water intake and output, and to circulate this water. We are talking about water, and not about all these sugary and alcoholic liquids which, on the contrary, hinder the redistribution of water in our body, if it is functioning properly and not if "my kidneys are too thin", to quote our brave soldier Gaston Ouvrard.

Nature is particularly imaginative in stressing us. Oxidative stress is probably the most permanent of our chronic stresses since we live in an atmosphere filled with oxygen, which is the oxidizing substance par excellence. Oxidative stress is also brought about by modern food or generated by exposure to ultraviolet rays (UV-A and UV-B), which generate the formation of ultra-oxidizing molecules. So, we protect ourselves and avoid prolonged exposure to the sun or in a tanning salon, which destroy elastic fibers and photosensitize collagen fibers. Unless you want to quickly acquire a mummy-like skin! Or, conversely, to acquire the thick, yellowish skin of solar elastosis when there is synthesis of elastic clusters that are also found in certain cancers or stretch marks. The pretty, well-tanned but very exposed skin of some surfers is not so beautiful when seen from the inside.

The mechanisms for fighting oxidative stress mobilize multiple resources in the body. Among these mechanisms are the internal error correction systems because the body makes mistakes all the time. Fortunately, it corrects them all the time. There are several quality control systems that eliminate anything that is misshapen at any time. A defect in this quality control system is the cause of moon disease in children who cannot repair the DNA of their skin exposed to the sun, or of certain mitochondrial diseases (Dubnikov et al. 2017). Sometimes, these systems are overwhelmed by excess stress, and this can result in cancers. Sometimes, they are hijacked, and this can result in permanent inflammation or autoimmunity. These are situations that can be found in chronic diseases such as Crohn's disease, ankylosing spondylitis, lupus or multiple sclerosis. For example, the number of cases of people with acquired cutis laxa (non-genetic) has increased significantly recently, either as a result of chronic inflammation possibly induced by pollution, or in connection with the development of myeloma or renal difficulties. Research is needed, especially since these symptoms resonate surprisingly with some side effects of certain vaccines.

4.4. Genetic stress

For children and adults affected by mutations, the priority is first to define the causes of the disease and then either to provide what is missing, to mitigate the negative effects with drugs, or to repair the DNA of the damaged gene (Figure 4.2).

Disabling mutations in children and the therapeutic wandering that often results are enormous sources of stress. It is obviously very troubling for the child. It will also be troubling for the parents who have to adapt to the change and manage the appearance of clinical signs of dysfunction as best they can. We have seen before that it is not possible, in most cases and at the present time, to provide what is needed to remedy the elastic capital defect even if research teams have duly taken up the task. The principle of medicinal DNA could be a lead. The concept has been around for a long time to replace the piece of DNA that carries the mutated part. Studies aimed at DNA re-characterization have focused and continue to focus on targeting specific tissues such as muscle, lung mucosa, the immune system or the retina. But the whole body cannot be targeted unless it is done directly in the embryo, as soon as the first cells are formed. New techniques could facilitate this action, but they will always raise considerable ethical questions, since the modifications would be transmitted to eventual descendants and we cannot exclude introducing new mutations by trying to repair one. The future will tell whether DNA edition technologies are safe and ethically acceptable.

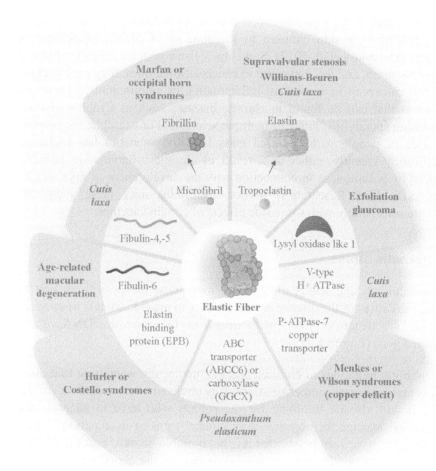

Figure 4.2. *Pathologies and syndromes of elastogenesis imperfecta. The figure illustrates the link between mutations in components required for elastic fiber assembly (shown in the Appendix) and genetic syndromes. Thus, certain syndromes clinically identified as supravalvular aortic stenosis, Williams-Beuren, Marfan, Wilson, Menkes, Hurler, Costello, occipital horn, as well as cutis laxa and pseudoxantoma elastica have been correlated in some cases with mutations in specific genes (see the Appendix for more details). These findings pave the way to therapeutic research*

More recently, in the wake of the search for a SARS-CoV-2 vaccine, the possibility of targeted RNA drug therapy has come out of nowhere. Admittedly, RNA was rather the uncooperative parent in the therapeutic toolbox, as RNA is very (very) fragile and difficult to control. The development of such a vaccine is therefore a real technological feat. It is not for nothing that some RNA vaccines need to be kept at an extremely cold temperature, in a freezer at −80°C. RNAs are the light intermediaries in the logistics between DNA and proteins. There are several types of RNAs. The best known are the last mile messengers that carry information to the protein production machinery. The most recently discovered RNAs have regulatory and action functions in the same way as proteins. As such, they are of interest to pharmacology and also to the subject of elasticity. In the case of the RNA vaccination of COVID-19, the choice was made to use an RNA as a messenger to force cells to build a decoy that will stimulate the immune response. This decoy is similar to a viral shell protein that has the very militant name of Spike. In all cases and especially in the fight against pandemics and cancers, the use of RNA drugs is a great advance in pharmacological research.

Another genetic stress relates to life expectancy and longevity, for those whose time is innately limited. There is an internal hourglass at the ends of our chromosomes. These ends, called telomeres, are necessary for the cell to copy the DNA of its own chromosomes and multiply. A defect of this DNA copy is encountered in some pathological situations. The most well known is progeria syndrome, which is related to a defect in the formation and function of cell nuclei (Jeanblanc 2014). The life expectancy of children with progeria is around 15–20 years. In a way, these child-adults have their lives accelerated. It was thought that telomeres would irreversibly shorten with each copy of the chromosomes, until the stage where copying would no longer be possible, inexorably marking the end of life. However, recent studies indicate that this shortening is partly reversible and that a good quality of life preserves the life of our telomeres. This discovery was the subject of a 2009 Nobel Prize dedicated to Elizabeth Blackburn, Carol Greider and Jack Szostak (Blackburn et al. 2017). This quality of life that preserves the length of our telomeres and thus our life is conditioned by food, environment and the mind but also by feelings since a high level of oxytocin, which is considered the hormone of love and social bonding, also has a beneficial effect on telomeres. Like a photocopier that regenerates its ink for love. In this context, and even if nothing has been demonstrated yet, it would be quite logical to ask the question of a possible link between elastic capital and the capital carried by the length of telomeres.

4.5. Stress and epigenetics

In the panoply of concepts and protocols that have agitated, agitate and will agitate the world of scientists for a long time, epigenetics occupies a prominent place. Globally, the term means "around genetics". One will find in particular the notion that the environment can put its mark and act on genetics, which is a real conceptual break with the "all-genetic" predetermination. Especially for those who encounter inequality when faced with the early appearance of wrinkles, the slowness of the healing process or susceptibility to disease. There are those who have wrinkles and those who do not! There are also those who are always sick and those who never are! Nature and nurture collide, without it being possible to really define the share of one and the other. At the interfaces of this discussion are epigenetics and the treatments that result from it.

Discussions of epigenetics touch on the shores of a tumultuous debate that took place in the early 19th century and that is resurfacing stronger than ever. The predominant theory of the early 19th century was that species were created once and for all by a creator: God. This theory of fixism was to be taken literally because it was written in the Bible. At the end of the 18th century, the physicist and botanist Jean-Baptiste Lamarck elaborated an adaptive theory of evolution of living beings under the influence of the environment. In other words, creation is not immutable; it evolves according to the environment. Lamarck was a bit ahead of his time and ended up in poverty and buried in a common grave. But the debate continued. Later, the biologist and biogeographer Alfred Russel Wallace on the one hand and the naturalist Charles Darwin on the other consecrated the selective theory of evolution. According to them, evolution is no longer adaptive but is driven by the selection of characters acquired through random mutations.

The debate was closed, except for some believers for whom fixism remained relevant. But the debate has just reopened as ideas of adaptation as a complement to gene selection are gaining momentum again (Penny 2015). A good example to illustrate this concept comes from an epidemiological study of Dutch people undergoing a severe famine orchestrated by the occupying Nazi forces at the end of World War II. Children born to women who became pregnant during this famine were found to be more susceptible to chronic disease than those of neighboring populations who did not experience the famine. But what was less expected was that the children of these children, in the next generation, had children who were also more likely to fall ill (Ekamper et al. 2014). Thus, there is indeed a transmission of information (the consequences of an action in childhood) that is statistically related to the environment (starvation). The information is brought about by an imprint on the genetic material that is transmitted to the descendants and can modify its expression (resistance to diseases, in this case). We then find the term epigenetics, when imprints or signals direct the genetics. And it is on these signaling mechanisms that we could have a therapeutic action. It is a situation of this kind that we have found with Romain Debret and our team in Lyon by analyzing the genome of children with cutis laxa syndrome or Williams-Beuren syndrome. The proven mutation of their elastin translates remotely into an imprint on other genes, which may worsen their disease and thus legitimize the search for a therapeutic bypass action (Debret et al. 2010).

These observations confirm that the gene and its mutations are not always the law and that it is possible to act on other levers than the single deficient protein. This changes the idea that everything is determined from birth by selfish genes and our heredity. Similarly, targeting epigenetic imprints is becoming a major issue in cancer treatment and behavioral prevention to optimize aging (Cruz-Jentoft et al. 2009). It should also be remembered that this history of imprinting on the genome occurs at each birth, through the transmission of information from the mother to the child. Yes, the mother can already make her epigenetic imprint in utero!

4.6. Pharmacology and stress

Many allopathic drugs target systems whose optimal functioning depends in part on the mechanical characteristics of tissues and organs. In reality, pharmacopoeia has few tools capable of addressing elastic capital if it is not backed by nutritional and behavioral efforts. Most drugs treat the consequences and not the causes of these innate or acquired pathologies. It manages infections and their consequences, external traumas (accidents) or internal traumas (stroke) or deleterious evolutions (cancer, diabetes, autoimmune diseases, etc.).

One of the targets of drugs for cardiovascular dysfunction is the balance of blood pressure. The renin–angiotensin–aldosterone system that mobilizes the kidneys, adrenal glands and the pituitary gland is very often targeted in the classical pharmacopoeia of hypertension and hypotension. SARS-CoV-2 does so with a timeliness that cannot leave one indifferent. An antihypertensive drug, losartan, has been considered a potential drug for the treatment of children with Marfan syndrome, related to a malformation of the fibril network underlying the elastic fibers. Losartan belongs to the family of drugs that target the so-called renin–angiotensin system. The hopes it brought were therefore logical. Unfortunately, as is often the case, clinical trials have not been sufficiently conclusive and its use as a drug for diseases related to a deficiency of the elastic capital seems less topical, although studies report its protective role for the vascular walls, which in itself would already be a good result (Isselbacher 2018).

To our knowledge, there is no drug that is considered effective and well tolerated for the treatment of fibrosis in general. Cross-linking inhibitors such as the active ingredients in the sweet peas that indirectly gave Francisco Goya the subject of his study might be good candidates, but they have many side effects. The same is true for treating scarring of affected tissue. Current treatments are used to slow the progression of fibrosis formation by treating the cause, when it is infectious and allows for it. This is not always the case, as current events demonstrate. A new avenue has just opened up by targeting the action of telomere formation, as if the rejuvenation of cells allows for better tissue evolution (Povedano et al. 2018).

Another approach is to consider the products of fiber degradation. These products participate in inflammation, amplify it and a vicious circle is established. This is the case when the products of the degradation and transformation of elastin accompany so-called scirrhous cancers (which are "hard"). We also sometimes find elastin deposits associated with the toxic deposits of Alzheimer's disease. These are manifestations similar to solar elastosis when the skin becomes modified by these pathological deposits of elastin. It is necessary to try to limit the degradation of these fibers, starting by avoiding sun abuse.

What could be described as a media flare-up about COVID-19 has shown how research follows leads, doubts, sometimes gives up, and then moves on to other leads. For example, new treatments are initiated in the case of cutis laxa acquired with a high pulmonary involvement and a high risk of morbidity, by taking up tracks already developed to fight against myeloma. Researchers are feeling their way but not giving up, in a permanent cycle of doubts, trials and discoveries that enrich a panoply of actions. This is also what those who want to manage their stress are doing, well aware that their body already has a lot to do and that it must be made easier by adopting habits and behaviors that do not add stress to stress, particularly through diet, pollution, sedentary lifestyle or psychological tension.

5

The Second Challenge for the Elastic Human: The Management of Food and Inputs

5.1. Introduction

It is a good idea to manage your health through diet, but we must consider food in the broadest sense, as an input into our bodies. These inputs will serve our metabolism, building and maintaining it via anabolism and producing its energy and activities via catabolism.

As we have previously focused on the management of disease and aging by preventing innate or acquired stress, we will consider in this chapter the theory of dietary restriction as a factor in longevity. We will then talk about sensible eating and its opposite, an unhealthy diet and all the toxic substances we can absorb that pollute our bodies. Since in reality, all this is not very entertaining and in the absence of a satisfactory pharmacopoeia to stimulate the formation of functional elastic fibers, we will begin with a botanical journey that offers multiple reasons for satisfaction and emotion. We will see how we can adopt the use of some plants and their families to manage our elastic capital with a deep therapeutic, visual, olfactory and gustatory satisfaction. However, the scientists in phytotherapy and chemistry will continue their investigations with, hopefully, a success as rewarding as the one that allowed the identification of many drugs from plants.

For a color version of all the figures in this chapter, see www.iste.co.uk/sommer/elasticity.zip.

5.2. Elastic capital and phytotherapy

Before searching for a molecule that would have a beneficial effect on the synthesis of elastic fibers, my initial training in plant physiology led me to consider an approach in phytotherapy. Plants are rooted fighters that must protect themselves from the attacks they have been undergoing for hundreds of millennia without any possibility of escape. They have armed themselves with chemical and mechanical weapons. They also know their own viruses whose action they try to delimit. Brownish spots on their leaves can testify to this fight. I began my research on the study of the defense of alfalfa against the spread of an RNA virus. Yes, alfalfa, similar to other plants, defends itself by trying to circumscribe the site of the viral infection. They do this by self-destructing the tissues surrounding the infectious attack, which causes these spots corresponding to the formation of a necrosis of the plant tissues to limit the invasion of the viruses. These properties have been used since the dawn of time by animals who know how to treat themselves with plants, and by humans who have not forgotten the heritage of their ancestors.

Science has corroborated the antiviral activities of plants against herpes, shingles, flu or hepatitis, for example. For this reason, extracts of astragalus, elecampane, lemon balm, St. John's wort, olive tree, oregano, licorice, marigold or elderberry are invoked. We can benefit from their weapons, always being careful that these weapons do not turn against us and adopting the usual precautions of the experts of official pharmacopoeia. But we shall return to the quest for the Holy Grail in this book, i.e., the molecule that would make it possible to rejuvenate by increasing elastic capital! If this molecule exists, it was not known previously, and it still does not exist today. Hence the importance of continuing research on plants that are real first aid (and poison!) kits. A review of the scientific literature had initially made it possible to take stock of the plants described to act on the consequences of aging in general, and of the tissues and organs presented here in particular (to summarize, see Simmonds et al. (2018) and always the list of official national pharmacopoeias). Thus, burdock, horsetail and bamboo are useful for building bones, tendons and ligaments. Horsetail, rich in silicon, is considered to be the plant that protects connective tissue. It is recommended during aging or light sprains. Arnica or calendula helps the healing of the skin. *Eucalyptus radiata*, marshmallow, licorice, plantain or

Scots pine is beneficial for the lungs. Butcher's-broom, bilberry, gingko, yarrow, garlic or sweet clover help the veins and arteries. Solidago and the olive tree can help to manage hypertension while valerian, lemon balm or hawthorn will calm any heart rhythm defects. Osteoarthritis and arthritis attacks can be soothed with savory, blackcurrant, angelica (the angel herb!), harpagophytum (the devil's claw!), wintergreen and lemon eucalyptus.

To cleanse the organs that regulate the important fluid balance of our body, we can choose between fumitory, dandelion, artichoke, fennel, black radish, orthosiphon or elderberry. We must also mention how the menstrual cycle can benefit from the help of yarrow, alchemilla, mugwort, feverfew or oregano. These plants may never be as successful as yew or Madagascar periwinkle, two plants that are toxic in origin but that have isolated some of the most effective anti-cancer molecules (taxol and vinblastine). However, they join the corpus of plants recognized for their benefits toward skin, digestive, urinary, biliary, renal or bronchial disorders, to mention only the treatments duly recognized by the registers of the pharmacopoeia. But no plant appeared satisfactory among this rich repertoire in the framework of our research on the activation of elasticity and the anti-wrinkle Holy Grail.

5.3. Elastic capital and dill

As there were no (in my opinion) plants known to effectively stimulate elastic capital, we started our own research which resulted in the identification of a dill extract. It may not seem like much, but this discovery was the culmination of a huge gamble. Indeed, the formation of functional elastic fibers is weak in adults, and no procedure was known to stimulate it, neither by pharmacology nor by traditional or ethnic medicine. This offered us little hope. Moreover, the first difficulty was that we did not have the shadow of a compound that could serve as a positive control for our experiments. Due to an efficient collaboration with colleagues from the Lyon-based company Coletica (which has since become BASF Beauty Care Solutions), we were able to test extracts from their product bank to look for a stimulation of an essential player in the formation of elastic fibers. This was LOXL1, which we have already highlighted for its role in fiber cross-linking. The hypothesis was based on the observation that this element is on the one hand essential for the assembly of elastic fibers, and that on the other hand its available quantity decreases during aging. It was therefore a gamble to be able to stimulate this LOXL1, and this would perhaps result in a rejuvenation of the skin by stimulating the formation of elastic fibers.

The consortium used long-term cultures of skin cells from postmenopausal women for this trial. This model was chosen because menopause is a devastating stage for elastic capital. Andropause is not great either, but it must be admitted that it is less rapid and less associated with an unfavorable stereotype. However, it should be noted that this last idea is evolving along with the male cosmetics market. But we will return to our experiment. A first group of plant extracts and chemical molecules was first selected on the basis of a stimulation of the famous LOXL1. And it was finally a dill extract that was selected because it was the one that most induced the formation of elastic fibers in our cell cultures. This activity was then validated on reconstructed skin synthesized by cultures of these same cells, which was achieved with the winning participation of the skin substitute laboratory of the Edouard Herriot Hospital in Lyon. Due to their model, our consortium effectively demonstrated a stimulation of the synthesis of elastic fibers and a protection against its degradation (Cenizo et al. 2006). This was costly and time-consuming work because the formation of these skins requires very delicate expertise brought about because of our collaboration between fundamental, clinical and applied research. The consortium of researchers then evolved to confirm in aged mice the results obtained on the reconstructed skin models in culture. With colleagues from a laboratory in Grenoble, we tested the effect of the extract on the cardiovascular system. The extract was administered in the drink of aged mice, i.e., about 2 years old. By a surely biased analogy, this would correspond to some 70 years for humans. Our colleagues, specialists in cardiovascular elasticity, observed that the arteries of the animals thus watered aged less than those of untreated animals and that their left ventricle, which does most of the work of pulsation and grew less with age, reflecting better elasticity (Fhayli et al. 2020).

To date, we have not identified a pure active ingredient from dill extract. It is never certain to isolate an active compound from a plant, but, as they say, research continues. In the meantime, we can recommend preparing a liquid extraction of dill seeds or crunching a few dill seeds ("organic") in the morning if you like aniseed flavors and if you follow your pharmacist's advice. First, it is very good for morning breath and digestion. Then, it links us to the great history which certified the value of dill through these words evoked in the Bible: "Woe to you, teachers of the law and Pharisees, you hypocrites! You give a tenth of your spices – mint, dill and cumin. But you have neglected the more important matters of the law – justice, mercy and faithfulness" (Matthew 23:23). This means that two plants of the umbelliferous family, dill and cumin, were valuable enough at the time to be given a tithe. This family is also called Apiaceae, but the name umbelliferae is still more poetic because it evokes strings of small flowers organized in umbels swirling on elegant branches.

Umbelliferae are a family of very diverse plants, some of which are well known by traditional medicines. This ethnopharmacology census is very good news for gourmets because the umbelliferae family is rich in many delicious spices. It also hosts hemlock, which differs from wild carrot only by the presence of purple spots on the stem, but which has perfectly opposite effects as the death of Socrates by hemlock poisoning testified (this is said for those who like to hike independently in search of wild carrot). As for dill, the cold extract comes from the seeds, which does not exclude the rest of the plant, but this point has not been tested to date (this is said for salmon lovers). The second plant extract we selected for our tests came from the leaves of fennel, another umbelliferous plant.

5.4. Epigenetics and marjoram

As I described earlier, the team I was leading was interested, under the leadership of Romain Debret, in epigenetic variations in children with a mutation affecting the formation of their elastic fibers. Let us recall that the term epigenetics means "around genetics" and suggests that the environment can leave its mark. We therefore looked for mechanisms involved in the cross-linking of these fibers that were sensitive to this type of imprinting in these children. As this turned out to be the case, we looked for and found another plant extract that acts on our favorite Trojan horse, the famous LOXL1. And this time, it was marjoram that was caught in the net, because of the collaboration with our colleagues at Beauty Care Solutions (BASF). The action of this garden oregano is located at the level of the epigenetic imprint, which limits the synthesis of LOXL1 in the cells of the dermis of donors (rather post-menopausal donors, for reasons already mentioned).

To date, there have been no preclinical trials with marjoram as there have been with dill, but its potential benefits add to a long list of recommendations for this Mediterranean plant of the Lamiaceae family, whose oil is often considered a natural antibiotic, to be used with caution, however, taking into account its potential caustic effect on the dermis or the liver at high doses. One must always consider that the effects of a plant extract are never simple and univocal. For example, here is a working hypothesis, concocted after preliminary tests on the activity of the LOXL1 gene in human skin fibroblasts treated with plant extracts (see the Appendix for a more complete explanation). Enough to feed several studies and to tire the brain!

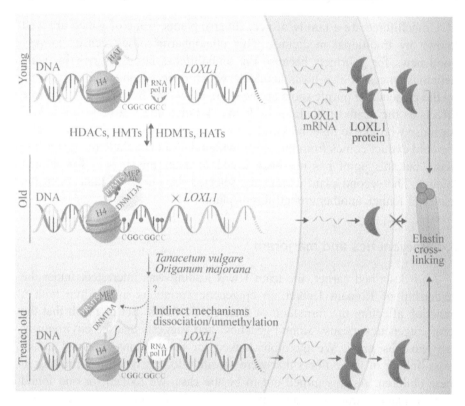

Figure 5.1. *Working hypothesis mechanism of LOXL1 gene promoter regulation during aging and in the presence of active ingredients. In young individuals, gene activity is facilitated by the epigenetic modification of histones (green circles). Histones are essential proteins in the nucleus of cells that regulate DNA compaction and therefore access to it. In young people, DNA is less compacted, more accessible and more easily readable. In older individuals, another modification of histones occurs which leads instead to more compaction and chemical modification of the DNA (gray circles) to prevent the reading of the LOXL1 gene. The plant extracts tested seem to act on this mechanism, and thus on the activity of our LOXL1 gene, the object of our study as an important molecule for the formation of elastic fibers (details in the Appendix)*

5.5. Adopting a plant

This presentation of phytotherapy expresses in fact a hidden desire, that of continuing pharmacological research on the extracts of plants and influencing everyone to adopt their own plant (always in agreement with a pharmacist). This is because the world of therapeutic plants constantly invites itself to our consideration and connects us to nature. Plants also stimulate our vision, sense of smell and proprioception. They can be found on balconies, on windowsills, in living rooms or in gardens. They are available at the herbalists or pharmacists who know their benefits and their side effects. They combine therapy with pleasure, with their vegetables and also their spices, this fabulous arsenal which covers up the least industrial foodstuff devoid of any attractive trait with regard to culinary desire, for us and our intestinal flora.

I personally have always consumed or used rosemary and lemon balm leaves and eucalyptus essential oils. I now add dill seeds and marjoram spices. And if there were only one left to choose from the list at the beginning of the chapter, it would be rosemary, which is proving excellent for my delicate liver. Rosemary was part of Hildegard of Bingen's selection, and its virtues have been recognized since Gallien, the father of Western pharmacopoeia. Many active principles of this common plant have been scientifically validated. According to Tao dietetics, the list of its properties concerns what Chinese medicine groups under the term "symptoms of cold humidity". It is even elevated to the rank of star under the name of Alecrim in the tradition of "Candomblé" (Brazilian voodoo) since it is associated with the most important of their gods, Oxala, the great Orishas. Even literature exposes it when Cervantes recommends it to create an ointment in *Don Quixote*, Alexandre Dumas makes it the basis of the recipe for miracle ointment passed on to D'Artagnan and Shakespeare introduces it in *Hamlet* to constitute a crown that guards the memory. "There's rosemary, that's for remembrance; pray you, love remember". This last tradition dates back to ancient Greece where rosemary was considered the guardian of memory.

So, here we are with beautiful plants, dill, marjoram and rosemary. The first one is part of the umbelliferae family, among which we will find ferula, which is used as an antispasmodic. Its stem was also used to correct naughty pupils, guilty of rebellion against authority. It is the opposite for angelica, this sweet herb of the angels, which is used against ENT complaints, flatulence, rheumatism and gout. Delicious coriander is recognized in prevention for its benefits toward the kidneys, diabetes and high blood pressure. The valuable caraway aids problems of digestion, and cumin is supposed to support healing. The indispensable fennel is diuretic and prevents biliary insufficiency and flatulence. The Mediterranean lovage was used in the Middle Ages as an aphrodisiac, but also much more concretely to alleviate menstrual and urinary disorders. As for the great hogweed, it became the muse of the art nouveau and was used as a model to carve the embellishments of chairs and woodworks. The ubiquitous carrot, chervil, celery, parsley and parsnip are indispensable culinary guests. And there is no need to present the resources of conviviality offered by green anise, especially when writing from Marseille.

The second set of plants, marjoram and rosemary, are from the Lamiaceae family. This family is well known for its therapeutic properties. But above all, we will agree on their culinary importance when its members are presented under their names of basil, lavender, marjoram or oregano, lemon balm, mint, rosemary, sage, savory or thyme. We should make a good choice and adopt one, several or all!

5.6. Elastic capital and protein restriction

This book began with a self-assessment, like a mirror of the time spent under the effects of gravity of our beautiful planet. Gravity that imprints a very visible sagging that is generally indicative of a loss of tissue elasticity. This observation bears witness to the reality of aging theories, and in turn, challenges the desire for the elixir of youth or rejuvenation. We have seen that elastin is considered to this day as one of the proteins, or even "the" protein, with the greatest longevity, with an estimated half-life of 70 years (Shapiro et al. 1991). We have seen that the body is very well equipped to fight against oxidation because it lives in a world of oxygen and elastic capital suffers particularly. Another theory of aging calls into question the joys and woes of nutrition. It is about caloric restriction, whose beneficial effect on longevity has been widely documented in animals. These studies have provided a wealth of advice on the practice of fasting, which represents the archetype of caloric restriction. In reality, in contrast to these results, the scientific follow-up of people who have adopted caloric restriction diets does not support the experiments carried out on animals. In humans, it is more a question of considering a qualitative effect of the type of proteins and amino acids ingested, rather than the overall caloric quantity. We should therefore speak of a protein restriction (Longo and Pelloso 2018).

The demonstration of the major role of protein restriction on aging deserves to be discussed in detail in the context of a story masterfully conducted by Valter Longo's team (Longo and Pelloso 2018). At the beginning of the story, we find people who have a genetic disease called Laron syndrome, some of whose manifestations may fall within the spectrum of mechanical defects. Indeed, these people are characterized by a small size, but they also stand out for their absence of civilizational pathologies such as cancer and diabetes, unlike other inhabitants of Ecuador where they live. A world specialist in aging at the University of Southern California, Longo demonstrated that Laron syndrome was due to a mutation of a gene that prevents the growth hormone's ability to function, hence dwarfism. He then discovered that if these people had fewer chronic diseases, it was because the regulations orchestrated by a certain hormone were less efficient. This hormone is insulin-like growth factor (IGF).

We have already encountered IGF1 in the section concerning the liver and glycemia. It is also the cousin of relaxin, which is present during pregnancy. It can act on the regulation of blood sugar by decreasing the level of circulating sugar. It protects the mechanics of blood vessels and tissues. But it should not be in excess because it is also associated with the stimulation of tumors. In short, it is necessary but not too much. A good way to regulate its production is to provide the body with glycine, an amino acid that is one of the building blocks of proteins. Glycine is even the major amino acid to constitute collagen and elastin (Sá et al. 2018). So, on the one hand, we have this fiber founding glycine, and on the other hand, this hormone for which there are studies proving that it is an incentive for the production of elastic fibers during growth. Here, we can see a link between a well thought-out diet and the formation of elastic fibers, the management of glycemia, the stimulation of growth and the increase of longevity in good health. This leads us to discuss the concept of a well-balanced diet.

5.7. Elastic capital and a reasoned diet

Many compilations praise the merits of specific diets. The claims are numerous. But few of them are based on scientific support and even fewer take into account the elastic system that is most often absent from nutritional considerations. However, we can unequivocally cite the holistic, scientifically documented and well-popularized recommendations of Valter Longo (Longo and Pelloso 2018) and David Servan-Schreiber (Servan-Schreiber and Dessert 2011). There are many others, but in summary, it is the variety and balance that prevails, which will be found in particular in the famous Mediterranean diet, with certain moments of light fasting or restriction as proposed by Valter Longo.

For the elastic system, the process is part of the logistics of synthesizing matrix fibers and obtaining a targeted (and non-toxic, like the calcium crystals that have surreptitiously crept into my knees) calcification. First, let us remember that vitamin C is essential, as we are reminded by the way the British navy managed to prevent scurvy in their ships. Scurvy wreaked havoc when the progress of the navy made it possible to envisage long-distance voyages. History has remembered the great discoveries or the stories of the great naval battles. For the record, it should also be remembered the loosening of teeth and bones, the damage to the skin and mucous membranes, foul breath, great fatigue, the decomposition of the heart and the death of thousands of sailors from all parts of the world. In those days, shipping companies anticipated (cynically) a loss of 50% of their human potential, until the British Navy introduced lemon into the food supply, in order to provide vitamin C preserved to remain functional.

Vitamins (B6, B9, B12, C, D, K1 and K2) are therefore at the forefront for enabling the synthesis of our fibers. Let us add copper, iron, zinc, magnesium, cobalt and sulfur derivatives to complete this first table. Note that all this is found in a reasoned and varied diet, without having to resort to expensive and often useless food supplements. We will also add slow sugars and lipids. Fatty acids omega 3 and omega 6 are very essential because the body does not produce them. They can be found in a well-balanced diet, but we must be careful because omega 3 is less present. In addition to these polyunsaturated fatty acids, we also recommend the queen of the Mediterranean diet, olive oil, which is rich in monounsaturated fatty acids that are very beneficial for the body. Proteins are always necessary because they are the building blocks of our body.

Once again, the Mediterranean diet prevails even if my Alsatian chauvinism obliges me to digress here on the merits of sauerkraut as a source of vitamin K (in fact K2), but to be honest, vitamins K1 and K2 can be found in other food sources such as the green leaves of many plants. In this regard, it should be noted that vitamin K inhibits pathological calcium binding to the blood vessel wall by interfering with the action of a protein called Gla (Bäck et al. 2018). It can therefore be said that vitamin K "sounds the death knell" for atherosclerosis and osteoporosis.

It is necessary to return for a moment to the question of proteins because they are equivalent neither for instructing elastic capital, nor for the other biological capitals. It is obvious that eating a steak or lentils does not offer the same nutritional contribution, quantitatively and especially qualitatively. Human proteins are composed of 21 interchangeable amino acids, eight of which are considered essential since the body does not manufacture them. This list can be extended in children or in certain physiological circumstances. Let us consider the two main proteins of fibers, elastin and collagen, to which I will add titin to complete the trilogy. Glycine, valine, proline and glutamine are the amino acids that dominate for their predominance in the three proteins and their role in elasticity. There are of course other necessary amino acids, but we will concentrate and look for these four treasures of our elastic capital in our food.

5.8. Glycine on the menu

Glycine is the simplest of the amino acids, probably one of the very first amino acids on earth. Glycine is integrated into collagen fibers at the rate of one-third of its composition and is very present in elastic fibers. It is therefore necessary for the growth of body tissues or for their repair. We have seen that it regulates the excess formation of IGF1, the armed wing of the growth hormone. It promotes growth and the healing of tissues, improves quality of sleep, memory, muscle mass and attenuates the symptoms of osteoarthritis and stress in case of chronic disease. It also lowers the level of glycation of the body. However, to always consider the negative effects, glycine, when provided in large excess, has been shown to stimulate in some cases the growth of tumor cells.

Although not essential since it is produced by the body, glycine is often in limited quantities in our modern diet. It is found in candies that are made with gelatin, which is obtained by cooking collagen from animal carcasses. But frankly, after having worked on dental caries, I cannot see myself recommending eating candy. Instead, we should look for glycine in vegetables and in the cooking of animal carcasses, meat or fish. Among vegetables, beet or quinoa are very rich in glycine. Glycine betaine, which is the beetroot version of glycine, is well known to athletes for increasing muscle mass and to food lovers, in the form of citrate in this case, for speeding up the third digestive period. Among meats, it is all the "leftovers" that are rich in glycine, i.e., ligaments, bones, skin and all that network of hard and rubbery fibers. It is therefore found in fish and meat carcasses. This brings us back to our origins, when humans arrived at the end of the feasts abandoned by the great predators and were left with only the remains. More prosaically, it is worthwhile reintroducing chicken broth and fish soup to the menu, especially when, for the latter, every scrap is used up, from nose to tail.

Valine, another essential amino acid, and its companions leucine and isoleucine, are also useful for the growth and repair of muscles and bones. They are found in abundance in lean meats, eggs, fish and vegetables. Similar to glycine, proline, described as non-essential, a term that current events encourage us to consider sparingly, is found in tissues sourced in collagen and elastin. Otherwise, proteins from soy, eggs, cheese, lean meats, fish and many grains, fruits and vegetables will do the trick. Fish and organs such as liver are rich sources of glutamine, in addition to eggs, dairy products, cabbage or parsley, which will help the intestinal flora and help manage irritable bowel syndrome.

Having said that, if you do not want to read any good diet book where the source of the amino acids is listed, the simplest, quickest and least expensive suggestion is to follow Mediterranean recipes rich in spices, and to sprinkle your food regularly with sprinkles of plant germ or brewer's yeast, which contain the eight essential amino acids.

To date, there is no documented diet to fight against fibrosis and the pathologies that stimulate it. A diet that does not promote inflammation is considered positive. On this point, we converge once again with the Mediterranean diet, with its low amounts of processed foods.

To finish, let us go back to the very beginning of our story and talk a little about breastfeeding because it is a current issue. The breast milk that is available when a woman first begins breastfeeding is rich in lactoferrin. This protein is synthesized by everyone, but it is found in quantity in colostrum. Among its properties, lactoferrin binds iron and therefore acts as an antibacterial by removing the iron that is necessary for bacterial growth. The mother thus offers additional protection to her child. In addition, lactoferrin stimulates bone growth. This and other widely documented benefits have resulted in a juicy market for lactoferrin substitutes for mothers. Whether these substitutions are reasonable or unreasonable is the subject of much debate, depending on the conditions of use. But the current issue that needs to be addressed is lactoferrin because it inhibits the entry of certain viruses, including the precursor to SARS-CoV-2. It thus becomes a potential weapon against COVID-19 (Kell et al. 2020). The future will tell whether the whole world will go back to breastfeeding to protect itself from the virus, especially since an excess of iron (ferritinemia) is a factor associated with an unfavorable evolution of the disease!

5.9. Elastic capital and an unreasonable diet

We have already presented the example of food toxicity in relation to the flour produced from sweet peas that was used to make bread during the Napoleonic invasion in Spain. Today, the hyper-sweet, meaty, oxidized and fatty food is our new invader that affects elastic capital. But first, to continue the discussion on glycine and longevity, we must introduce the example of glyphosate to evoke the difficulty of assigning to a molecule a long-term toxicity.

In general, the identification of toxicity is defined over very short periods of time, from a few hours to a few weeks. The effect on cell multiplication or fate is essentially evaluated, which generally concerns the process of carcinogenesis. But this is not sufficient to assess impacts such as degeneration of bones, muscles, nerves, vessels and tissues. The study of glyphosate is an example of the difficulty in establishing the potential long-term toxicity of a product. Glyphosate is chemically a phosphate-linked glycine, which raises the possibility that it may replace the natural glycine in our tissues, particularly in collagen and elastin. Glyphosate is also capable of capturing metals such as copper, essential for the stage of cross-linking of our fibers. The consequences of a defect in the use of copper are known because this defect induces Menkes syndrome, which affects children with neurological, growth and coordination disorders. These children also suffer from a flagrant deficit of collagen and elastic fibers. All these observations converge to feed the polemic already in progress on the short-term toxicity of glyphosate, although to date there is a lack of long-term studies (Mesnage and Antoniou 2017).

The verification of long-term safety on the functioning of body mechanics is emblematic of the procedures to be developed for any product resulting from a technological and chemical innovation. But these procedures generate a significant human and financial cost and require interdisciplinary knowledge that is rarely assembled in long-term toxicity studies. For example, it was necessary to feed aged mice for several months to validate the positive effect of dill extract on cardiovascular mechanics, by combining several fields of expertise. And then we talk about prevention, the field that has become so poor in financial resources in our emergency pharmacology.

The toxic effect of an excess of sugars in the long term is however perfectly documented. Glycation is at work. Diabetes is a destructive weapon, and its executioner is sugar. The presence of glycated hemoglobin, which is well known to diabetics, traces its action. It is therefore preferable to limit sugar consumption to avoid "caramelization" and the subsequent deterioration of organs, whether these sugars are pure, simple and quickly assimilated (white sugar and all industrial sources of sugar) or in the form of slow and complex sugars (cereals, pasta). This would avoid, for example, preparing with certainty the future chronic diseases of children such as soda disease and promoting the multiplication of dental caries.

There is another excess that is definitely toxic for elastic capital, which is less known than the excess of fats and sugar. There is a disease called hyperhomocysteinemia (de jaeger et al. 2010), which is not often talked about, perhaps because of its long name! And yet the disease affects the cardiovascular system as well as contributes to certain dementias and even Alzheimer's disease or Parkinson's disease (Kumar et al. 2017). The disease is due to an excess of homocysteine, which is an amino acid essential to health in normal conditions. Its excess can come from a lack of B vitamins (12, 9, 6) or glycine in the diet or, conversely, from an excess of sulfurous amino acids. This can be translated by avoiding excess red meat, which is both rich in methionine (another sulfurous amino acid) and low in glycine. While too much meat is detrimental, white meat, fish flesh and vegetables can be favored.

There is an open question about lactose in the diet, since most adults do not have the enzymes to digest it. The debate also exists in the small world of researchers working on elastic fibers because there is competition between elastin and lactose. As a prelude, it should be noted that the observation of this competition has only been made on cells in culture. The elastin produced is taken out of the cell by a shuttle protein, which releases it onto the weft of the external fibers. This is where the enzymes that work on the cross-linking of the fibers (the "LOX") come into play. The lactose uses the same shuttle to enter or leave the cell and thus competes with elastin (Qa'aty et al. 2015). Formally, this would seem to indicate that lactose, and thus milk, might be involved in the formation of elastic fibers in humans. This is an unresolved question, one that is not close to being resolved, due to the complexity of the models that need to be implemented to resolve this hypothesis.

To complete the current debates on food, it is necessary to talk a little about gluten. The two major protein components of gluten provide the elasticity and extensibility properties apparently required by the large-scale preparation of industrial products. But it is its ubiquitous use as an additive in food products that induces negative reactions for the elastic human. The best known form is celiac disease, which results in a complete intolerance in the intestinal wall. The prevalence of this disease is increasing, with an increased risk in women and diabetics. This observation promotes the adoption of a gluten-free diet for intolerant people, or with little industrial gluten for sensitive people. Without making such a strong causal link, the acquired syndrome of cutis laxa may also be linked to gluten and celiac disease. Whether due to improved detection or a latent increase, there are indeed more and more cases of acquired cutis laxa whose causes suggest the malformation of antibodies that clump in the kidneys. A gluten-free diet is recommended for these people, which can be found in a Mediterranean diet (but with no gluten-containing pasta and no couscous).

To finish this section, we will mention obesity, one of the morbidity factors frequently evoked during viral infections. We are faced with two pandemics, viral and metabolic, whose consequences appear to be cumulative. Once again, we will limit ourselves to the question of elasticity in this complex subject. Fat accumulation can have different effects. The first effect concerns the diffusion of gases and fluids in the body. The circulation of oxygen, carbon dioxide and nutrients between the blood (in the arteries or veins) and the tissues (including the muscles) is not facilitated by these local fat deposits. These are the Charybdis and Scylla that disrupt the path of our gases, as we saw in Chapter 2. Second, the elasticity of the vessels is properly limited by the chemical changes caused in the vessel walls and by a high glycemic or lipid level in the blood. The slow flow of lymph will also become entangled. High blood pressure often associated with obesity will favor the entry of the virus by the reinforced presence of the famous ACE2. Shortness of breath, reduced respiratory capacity and frequent sleep apnea will facilitate the virus' hunting ground, especially in the advanced stages of pulmonary fibrosis. Another aspect is the natural hardening of this fatty tissue. In a way, this tissue becomes diseased, inefficient and hardens by the formation of an internal reticulation that will make it even more difficult to return to a body balance. It is therefore essential to take into account this state of obesity, which is not only linked to a problem of diet but also to particularly complex genetic, epigenetic, metabolic, microbiotic, psychological and societal factors.

5.10. Elastic capital and pollution

The bronchial tubes are the front line in the body's management of pollution. This results in local inflammation, following an infection, or chemical degradation, by oxidation (cigarettes, etc.). This can destroy the elastic fibers under the walls of the alveoli, which results in a decrease of respiratory capacity, sometimes leading to emphysema. Inspiration and even more so expiration become difficult because the elasticity of the walls is altered (increased compliance). The compensation by a contribution of new collagen fibers can lead to a rigidification and result in fibrosis. The exchange of oxygen and carbon dioxide with the subpulmonary veins and arteries becomes difficult, leading to respiratory failure. When the pressure balance (osmotic and oncotic) is altered and the heart is less efficient, the pulmonary tissues can compensate by filling with water, which induces pulmonary edema. Convection through the vascular system is less efficient, with less transfer between the alveoli and the arteries and veins. In addition, the mucus (surfactant) that lines the outer walls of the alveoli may be altered and lose its properties. This is the case, for example, of fine particle pollution. The diffusion of oxygen in the body is then reduced and the respiratory capacity is affected (Figure 5.2).

Several sources of pollution have a notorious effect on elastic capital, especially on the voice and breathing. The smoker's hoarse voice is the first evocation that comes to mind when talking about the effect of the local environment on elastic tissue. In this case, it is the toxic effect of oxidants generated by a cigarette's smoke on the fibers of the vocal cords. We can add pathologies such as bronchiolitis, asthma or emphysema, linked to the interference of microparticles with breathing. These microparticles can be associated with the mucus lining the walls of the pulmonary alveoli. The finer the particles, the more they will go to the bottom of the pulmonary alveoli and affect tissue function and elasticity. This observation, absolutely undeniable (Manisalidis et al. 2020), should signal the final term of the use of situations and tools dispersing these fine and especially ultrafine particles in our immediate environment.

Another source of pollution comes from the transformation process of natural lipids called the hydrogenation of fats. In nature, it is ruminants that produce these hydrogenated fatty acids that are called trans fats, while humans do not. The (industrial!) interest of these hydrogenated oils is to modify the texture of food and to increase their conservation. But for human beings, it is absolutely essential to avoid these trans fatty acids because they increase the risk of atherosclerosis.

There are several mechanical risks that can alter the proper functioning of arteries. We have already seen that the necessary elasticity of the vascular walls can be altered by an excess of sugar (glycation), calcium, or homocysteine. We will add the fatty acids that love to stick to elastin, which is itself very fond of fat. The walls of the vessels stiffen, the heart pumps more, oxygenation is less efficient, the tissues suffer from a lack of energy and the body tires. The interior of the vessels can also become partially blocked with the formation of atheromatous plaque rich in cholesterol on the artery walls. Trans fatty acids stimulate the formation of low-density cholesterol (LDL), which is considered bad cholesterol because it contributes to the formation of plaque. The vascular walls in front of the atheromatous plaque become thinner and lose their elasticity. The consequences can be disastrous, with stroke, infarction, thrombus, ischemia or embolism. So, we might also prevent all this by looking carefully at the quality of the oils used in the food we ingest.

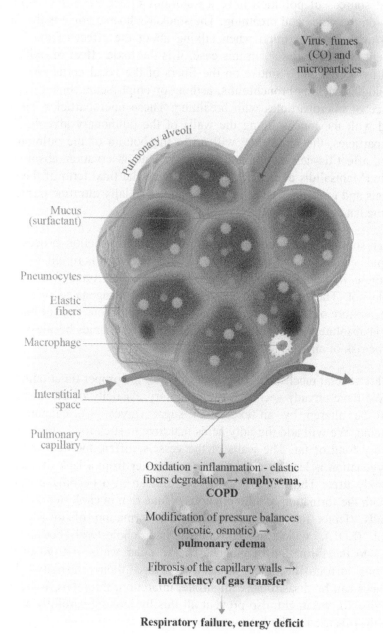

Figure 5.2. *Impairment of the elasticity of the pulmonary alveoli*

Another recurring consequences of pollution is fatigue. We have already mentioned the major role of mitochondria in energy management. Remember, they are these little energy factories nestled in our cells to produce ATP, which should arouse our deepest respect. Unfortunately, it is not certain that this attention is shared by the industry because they suffer from the side effects of treatments with fungicides and pesticides or from the excess of heavy metals. Let us take the example of copper, which we must not ignore; otherwise, our elastic and cognitive systems will collapse. But on the other hand, its intensive use by industry makes it too present. The effect on the mitochondria of all these pollutants results in a deficiency of energy production (ATP), especially in muscle or nerve tissue (e.g. in mitochondrial myopathies). Our mitochondria are the subject of studies to understand the development of cancers (Kenny et al. 2019), neuromuscular diseases (such as Parkinson's disease or Alzheimer's disease) or diseases still not understood such as chronic fatigue syndrome, fibromyalgia or muscular dystrophies. If we add to this the irreversible muscle wasting linked to age and associated with a decrease in the number of mitochondria, we can bet that the elderly will be all the more sensitive to pollution and diseases (Bhatti et al. 2017).

Plastic swords of Damocles hang over our heads. While viruses, bacteria and parasites are responsible for often fatal epidemics, other epidemics are at work in a sneaky but constantly orchestrated way. The mind-boggling quantities of nanoparticles and pollutants derived from these plastics are such that they can initiate a biological disaster, either directly by ingestion of pre-packaged food or indirectly by their accumulation in the food chain. We can enumerate the catalog of industrial inventiveness to generate poisons, between phthalates, bisphenol, polycarbonates and all other forms of plastics and surface coating of packaging and containers; even baby bottles were covered with them for a while, sure to generate a problem at the beginning of the lives of little humans.

To summarize, elastic capital is very dependent on food. The attention that we pay to it in our life course conditions to a great extent our energy and quality of life. Dietary deficits or excesses on the one hand, as well as the drifts of excessive industrialization on the other hand, have significant negative repercussions on the management of this capital. It took all the pressure of a tiny virus to finally make us realize that the air we breathe, the food we eat and the water we drink are our sources of life that must not be polluted in any way. And when we say "we", this includes the tiny microscopic world that supports us.

6

The Third Challenge for the Elastic Human: Successful Life through Movement

Figure 6.1. *The elasticity of movement. Elasticity is ubiquitous in movement. Its iconography is illustrated by lines of force and tension, as well as perspectives and vanishing lines. Moreover, elasticity oscillates between a hyper-elasticity where the functions are multidirectional and plasticity where a certain orientation is de rigueur. To represent all these particularities of elasticity in movement, I was inspired by Vieira de la Silva, whose compositions are illustrated by a meticulous representation of perspectives, tensions, lines of force and segmentation by finite elements of systems such as libraries, battles or games*

For a color version of all the figures in this chapter, see www.iste.co.uk/sommer/elasticity.zip.

6.1. Introduction

There is no shortage of scientific essays on movement. One can find innumerable treatises dedicated to the motor faculties, between the body material and the mind. To go further, I can recommend the texts of Alain Berthoz, master in the explanation of the "sense of movement" (Berthoz 2013). In the midst of this very interesting information and practices, this chapter aims to shed light on the notion of elasticity during movement, highlighting the essential notion that movement is the most recognized means to this day to slow down the inexorable decrease of elastic capital.

Movement and freedom of movement characterize many aspects of human life. The evolution of our ability to move was accompanied by the optimization of the heart, the selection of bipedalism, and a singular ability to perspire that allowed us to run fast and long, which certainly offered a selective advantage to the poor, very puny beings that we were. Evolution also brought us a gripping hand and haptic sensations, allowing us to shape and use tools for survival, domination or creation. The mastery of movement has enriched the cognitive faculties to measure, control and improve our actions. The evolution of techniques has then allowed us to go beyond our mechanical and sensory limits. It is thus a considerable subject located between our natural limits and the mastery of techniques allowing the expansion of our capacities.

Everything in the body moves, even the teeth, most of which are surrounded by a periodontal ligament with proprioceptive powers. In a somewhat trivial way, elasticity offers to movement what the tire offers to the wheel and to the engine, i.e., the tool that facilitates and allows movement in a pleasant and efficient way. Try riding a bicycle without tires and on rims! But it is much more than that. You certainly have to consider the tire and the wheel, but also the balance and the sensory organs as well as the notions of dynamics and kinesis. Before we get on the bicycle as an anti-depression tool, we will discuss some of the mechanisms that regulate the alchemy of movement. Then, we will consider two critical phases of our life: when our body is building up and is constrained in its uterine niche and when our natural cardiac pacemaker inserted in its elastic niche imposes our rhythm of life. We will then examine the creation of associative reference frames between the perception, interpretation and mobilization of positive and negative control systems. As there is nothing better to move than to enjoy it, we will try to flatter our six senses in the practice of physical activity.

6.2. The alchemy of movement

In this book, we have encountered several chemical molecules that are essential for movement. There are those that are used to produce the necessary energy, like ATP, oxygen and glucose, and there are those that control, activate or inhibit actions, such as adrenaline from the sympathetic system and acetylcholine from the parasympathetic system. As a reminder, because it is necessary to help the memory when faced with this avalanche of data, the sympathetic system is exciting and vasoconstrictive, while the parasympathetic system is calming and vasodilatory.

About 60 other chemicals shape our moods, behaviors and responses to environmental situations. If we are only interested in stars, these are the ones that give us goosebumps or make us alert (noradrenaline) or react to stress (cortisol), reward us when the music is good (dopamine), stimulate mother–child bonding and promote social connection (oxytocin), positively affect our mood (serotonin), require sunlight to adjust to jet lag (melatonin) or make us addicted to sports (endorphins). We can specify that adrenaline prepares us for stress, love, fight or flight by accelerating the heart rate in order to improve oxygenation of all the muscles of the body; acetylcholine, which immobilizes us in the face of danger, regulates breathing and slows down the heart rate; GABA slows down the transmission of nerve signals and provides calm and relaxation; oxytocin acts on the cervix during childbirth and on the mammary glands to initiate breastfeeding; dopamine is essential for muscle movement (Parkinson's disease is linked to a lack of this), the secretion of the growth hormone or the regulation of pain and depression. All these mediators are stimulated by physical activity and produced by a varied diet and by the balanced and synergistic contribution of different organs.

These examples reinforce the link between movement, body alchemy, tone and mood. And vice versa! All we have to do is adopt the right stimulating practices in our life paths! But we must also know how to relax and respect the brain's rest. This indefatigable hero recovers in part during sleep and gives itself a great cleansing, when the slow flow of the cerebral lymphatic system sends to the kidneys the maximum amount of waste produced by its hyperactivity. It breathes a little during the breaks and takes advantage of them to memorize and classify the day's information or synthesize a certain number of hormones, such as melatonin. It anticipates the future and creates response loops that will shorten its reaction time. Unbeknownst to us, it will form unexpected associations that will shape the ideas of the early morning, thus concretizing the adage of "sleeping on it". Everyone has experienced the benefits of sleep regarding clear-sightedness. As for the followers of meditation, there is no need to tell them about the benefit of this practice on the elasticity of the body and the mind. This convergence of brain and body activities is the basis of elastic thinking exposed by Leonard Mlodinow in his book *Elastic* where he explains the mechanisms that allow the elasticity of the mind ("flexible thinking") (Mlodinow 2018).

Great fatigue is associated with a deregulation of these numerous life regulators. In addition to the tissue alterations we have seen (bronchial tubes, vessels, heart), this fatigue can be due to anemia, which occurs due to a lack of iron that will fix oxygen. It can also be due to the deregulation of blood sugar or the thyroid, causing fatigue and cold. But it can also be a great fatigue of our mitochondrial batteries. One of the ways to increase their efficiency is to increase their number through sports. The muscles are stimulated and produce more mitochondria and therefore more energy (Ruegsegger and Booth 2018). This is a good way to protect against aging and against diseases (Moreno Fernández-Ayala et al. 2020). In other words, to age well and resist the bugs that bother us, let us move and mobilize our mitochondria.

6.3. Elasticity at both ends of life

We have seen in this book how much elasticity obeys the common laws of mechanics and biology. However, there are specificities according to the period of our lives. Growth is the period of structuring, and the rest of life is the period of strengthening or decline. We can therefore analyze the question from both ends of life and start with growth, in and ex utero when the elastic capital is built up between the first months of life and the end of growth. At the beginning is the movement of the fetus. There are situations where the fetal movement is altered or blocked. Bad positioning can induce a seizure of the joints. This arthrogryposis results in deficient muscle formation, fibrous retractions around the joints and nerve alterations. It will need to be treated with orthopedics and physical therapy at birth. At birth, the body abandons the gentle weightlessness of the amniotic fluid and is subjected to gravity, the air distends the lungs, and the blood circulation becomes a closed circuit. With so many changes outside and inside the body, it is not surprising that the perinatal period corresponds to the peak of elastic fiber synthesis. Childhood and adolescence will be the periods when physical activity will be crucial to building up the elastic capital; this is a period when the stimulation of fiber formation remains very active under the pressure of growth hormone and IGF1. Considering appropriate physical exercise during young adult training is not an exercise in style, it is a necessity to shape the elastic body.

At the other end of the story, muscle wasting, degeneration of elastic tissue and sensorimotor alteration will combine to sculpt the image of old age, more or less delayed depending on our efforts to maintain the body. But there are data that are relatively independent of our efforts. These are heart rate and its corollary, aerobic capacity. The programmed and irreversible decrease of the heart rate completes the theories of aging that we have already seen (Gluck et al. 2017). The loss of aerobic decline can be countered with appropriate tonicardiac maintenance. This involves training the so-called VO_2 max, which is the maximum volume of oxygen the body can consume. But it is much more difficult to manage the decrease in heart rate. This rhythm oscillates between 60 and 90 bpm in adults, with a decrease of 0.6–0.8 bpm per year from the end of the growth phase. It is maintained by cells that form the biological pacemaker. It is on the rhythm of electrical stimulation by these cardiac cells that life ultimately depends. We owe them infinite respect, even if we can now replace them with an electronic pacemaker. The cells of the biological pacemaker are connected to the sympathetic and parasympathetic systems. The heart is de-innervated and taken out of context, and it beats at around 100–110 bpm. The vagus nerve helps to slow down the rhythm, which strongly suggests doing all those activities that stimulate the vagus nerve, such as singing, walking, yoga, qigong or meditation.

We can see in the cardiac example how movement is essential to life, and how this movement must be perfectly orchestrated by the neurological network, with the good news that we can act through our sympathetic and parasympathetic systems. The heart has walls and valves that are duly elastic; otherwise, it would not function. Moreover, it does not function well as soon as it is subject to fibrosis or arrhythmia. But it is less known that our biological pacemaker is comfortably positioned in an elastic niche, at least much more elastic than the other walls of the heart (Gluck et al. 2017). Since nature rarely selects only what has an advantage, it would make sense that the function of this sino-atrial node benefits from the elastic properties of its niche, as do the other tissues involved. This could perhaps solve part of the mystery of the differences observed between individuals because we are not all equal either in our heart rate or in its aging. To date, knowledge is lacking to verify this hypothesis, although the establishment of fibrosis is known to be detrimental to our natural pacemaker (Csepe et al. 2015).

6.4. The reference frames of motion

Many practices exist to maintain or improve body mechanics. Some of them involve risks, which may have led to Winston Churchill's iconic "whisky, cigars, and low sports" to explain the secrets of his longevity (he died at 80, which was notable in his day). One of the most significant causes of injury is the repetition of a traumatic gesture, which can lead to what is known as musculoskeletal disorders. The cynicism of this is illustrated by Charlie Chaplin's *Modern Times*, where he uses the rhythm of the assembly line to work. In another register of society, the violinist will have a hard time avoiding pain, twisted around their beautiful tool. However, the use of the mouse by the *"homo informaticus"* will inflame many ligaments and tendons. Therefore, there is a real need to identify inappropriate postures to avoid or reduce this stress. It is necessary to identify the constraints imposed on the organism, with tensions that are too strong to the point of almost breaking, inflammations with calcification of ligaments, or repeated mini-traumas such as on the intervertebral discs or on the knees and feet. These biomechanical considerations should always be taken into account to ensure that physical practices do not permanently degrade the elastic capital of the body, especially in young children engaged in sports competitions or people who already have physical difficulty.

We have seen the stress imposed by inappropriate, repetitive and even violent movements. Conversely, a sedentary lifestyle, whether imposed or voluntary, results more or less rapidly in a loss of elastic capital, muscle wasting and sensorimotor degeneration. Not to mention the pernicious effects on the mind, sleep, cognitive faculties and even life expectancy through the regulation of telomere length. This is the very vicious circle faced by people suffering from painful and disabling chronic diseases.

Many circumstances are associated with movement, whether it is the gesture of the Paleolithic hunter-gatherer, that of the modern soldier armed with technology or that of the pétanque player. In all cases, the interface between neurons-muscles-joints-perceptions is mobilized. This interface is fed by experience, both environmental and personal. The gesture of the pétanque player and that of the soldier have a target, even if they apparently have little in common, although sometimes the adrenalin level of the Provençal pétanque player can join that of the warrior. As we have introduced, the process of association between their targets and their movements will involve the creation of associative reference frames between perception, interpretation, the mobilization of positive control systems (this is the "I shoot" of the pétanque player) or inhibition ("I point or I break everything") and the final action (shooting or pointing). The retinas of the bowler and the hunter will print similar information, i.e., light impulses. These signals will be transmitted by specific circuits to nodes of neurons (hubs), connected to a base of gestural analogies ("I have a gun in my hands" or "I have a pétanque ball"), decision loops ("I run" if I play Lyonnaise-style pétanque or "I stay on my feet" if I play it the Marseillaise style). The brain will have to take into account the electrical signals of nerve feedback toward the muscles with times of displacement which will be a function of the distance (less than 1 m for the fingers, almost 2 m for the feet). It is this process of building a reference frame that will allow the brain to anticipate and become creative in the action, all the more so if the training has consolidated the reference frame. To come back to my pétanque player, his body's reference frame will evolve according to his muscular agility, his sensory capacities and elastic capital, without forgetting of course some distortions linked to the sometimes alcoholic practices of conviviality.

To illustrate the link between anticipation – action – elasticity, we can take the example of mastering the cup-and-ball. It is a question of throwing a ball (held by its string) on the rod adjusted to the diameter of the ball hole. The sensory primacy is given to the synergy between vision and proprioception, with a high awareness of the different degrees of freedom that the string and then the ball could take. The exercise is even more interesting (in terms of amplitude, not success) when using a slightly elastic string. But it is better to perform this exercise away from any object or person to avoid accidents!

The brain is therefore capable of performing multiple tasks at the same time, which is evidence of its high flexibility (Mlodinow 2018). There are the automatic tasks, often delegated to the local level, but always linked to all sorts of sensors to monitor their functioning, body position in space, and physical and metabolic balances in real time. There are the conscious background tasks, such as standing and looking around. There are focused tasks, such as cooking. There are the random tasks, coming out of nowhere, such as adding an unexpected ingredient to the tried and true recipe because all of a sudden "you remember reading it in a book". There are the backup tasks, such as retrieving the wrongly positioned pan *in extremis*. There are inhibiting tasks, such as not picking up the hot pot with our hands. So, there are automatic and thoughtful actions. There are also senseless, uncontrolled actions that result from unexpected connections and that project into the unknown and unexplored. The brain manages all this, without necessarily an exogenous input of information. The two cerebral hemispheres work together, and although it is not absolute, we may consider that logic and reason are rather orchestrated by the left hemisphere, while the sources of creativity rather find their origin in the right hemisphere (Berthoz 2020).

6.5. Flattering the view and looking good

It is a fact that the increase in the availability of stylish and practical sportswear has increased physical practices. Visuals matter, for yourself and others. Bipedalism, from leisurely walking to running, is a great example of an exercise that can flatter the visual and offer us a reward. There are several structural systems in the body that allow for a pleasant and efficient gait. These include the feet and the head carriage. Their optimal use gives the impression of being free of gravity while fighting against bone demineralization. There is also an old theory that mineralization is stimulated by the production of electricity in the bones. The latter would behave like a material with piezoelectric capacity. This means that they are capable of producing electricity under mechanical stress. The theory has found many echoes in practice, for example, in the intervertebral discs

(Poillot et al. 2020), although it does not need to be demonstrated everywhere to know that each well applied step brings its spark of life.

While we wait for our turn for a weightless ride, let us start with the foot, which suffers a lot from our lack of interest in mechanical things and our fascination for fashion conventions. It is however the major element that gives us allure and guarantees a beautiful longevity without pain or constraint. In fact, it is a pure marvel, always in great demand. It is the guarantor of our bipedalism with its 26 bones, 16 joints, 107 ligaments and 7,200 nerve endings. It is the basis of posture and movement. Having practiced long-distance walking, Nordic walking and running, I have often been struck by how little consideration is given to biomechanics in the choice of shoes and the way one uses one's feet, which leads to all sorts of ailments when the foot meets the ground in an awkward way or is too tightly packed in the shoe. It seems that everyone walks as they can, rather than as they should. Walking is not such an easy exercise; each step mobilizes many muscles that are either contracted or stretched (which solicits muscle elasticity). The balance between the steps is part of a relatively complex gait cycle that involves the synchronized coordination of the bones/tendons/ligaments/fascias of the feet, ankles, knees, hips, shoulders, arms and spine components. The amplification of this gesture is achieved in practice with sticks, like so called Nordic walking. A real education according to one's personal biomechanics would be judicious, like the one that recommends the adoption of a dynamic walk. This implies a mastery of a gentle heel strike on the ground, the rolling of the foot and the propulsion with the support of the big toe to offer maximum efficiency and longevity for the entire body structure. This includes the foot, of course, but also all the other motor and elastic structures. So let us take great care of our feet, with intelligence, let us care for them, wrap them up with cling film, and put on wide shoes when conventions allow it so as not to stress our toes. The large feet of Alberto Giacometti's *Walking Man* are a perfect representation of this beautiful look. For those who abhor walking, or simply do not feel capable of strenuous exercise, there are a host of very simple, affordable and undemanding postures to get your foot on the ladder.

A second structure that I want to take as an example concerns the hyoid bone and its attachments. The hyoid bone is well known by those who follow detective stories with their cohort of forensic scientists who will know whether the fracture of the hyoid bone is evidence of strangulation or not. This bone, which solidifies at the end of the growth phase, has the particularity of being the only one in the body that is not connected to any other bone. On the other hand, it has links with the anterior cartilages of the neck and jaw through attachments to the muscles the tongue, the pharynx and the floor of the mouth and the front part of the neck. Placed under the tongue, it can be felt when pressure is applied to the lower jaw with the fingers. It bathes in a universe of muscles, ligaments and fascias. These are collagen fibers associated with some elastic fibers that run through the body. The fascias were somewhat neglected before becoming important again because their capacity of support, extensibility and elasticity appeared important. The fascias correspond to a real system of fibers and tissues that support and accompany the mobility of all the elements of the body. They can be found under the skin, around the muscles or the organs. They form stabilizing links between the different sections of the body. The decrease in their mobility is concomitant with immobilization, for example, following a fracture. This is another network that can get damaged if not used (Bordoni and Zanier 2014)!

In total, the vertebrae, the hyoid, the fascias, the ligaments and the muscles make up a system that reinforces the work of the spinal column and constantly supports the vision. This is no small thing when you know that a single vertebra supports the 5 kg (approximately) of the head and that this weight can be increased when the head is not vertical. Atlas carried the world and the heavens on his shoulders for eternity according to Greek mythology, and the first Atlas vertebra carries our head on our shoulders for the duration of our life. This system that participates in the orientation of the head and the vision deserves our consideration. This is why we can perform stretching exercises of the fascias and muscular relaxation to relax the accumulated tension. These exercises can be done with postures such as neck stretching, twisting or the yogic warrior posture, just mentioned to encourage simple and beneficial practices, to be done according to one's means. This can only be good after long periods in front of a screen where our vision, obsessed by the reduced surface of our screens, forces our head to adopt the famous turtle position. What Greek mythology had not foreseen.

The appearance is also a story of the eyes. We will not speak here of the color or of the ellipse of the eyes, but of vision itself. The connection between vision, proprioception and movement is extremely strong. Many of these pieces of information are convergent and act in synergy or in relay. An example of synergy between vision and action is studied by teams from Aix-Marseille University and Sorbonne University to mobilize the phantom limb to guide a prosthetic arm (Touillet et al. 2018). The brain indeed retains the image of the missing limb imprinted by vision before amputation. For three-quarters of amputees, this imprint translates into pain that we must seek to annihilate. The idea is instead to reallocate the brain's resources to command the muscles of the stump (if it remains) in order to direct a connected prosthesis, which requires neither surgery nor special training.

The beauty of the gait and the ability to maintain a good posture are also conditioned by a sixth sense, proprioception, which we will now discover.

6.6. Feeling your proprioception to improve your posture

Proprioception allows us to define our position in space, i.e., in a three-dimensional space underpinned by gravity. The body constantly captures all the information necessary to situate itself in time and space, with breathing as a modulator. Vision and sometimes hearing bring their convergent information. We previously mentioned the case of someone who has lost her proprioceptive links and who, in the dark and therefore without visual aid, feels like a floating brain, without knowing where her legs, arms and body are. In her case, vision predominantly compensates for the absence of proprioception. This is called vicariousness, when one sense compensates for the failure of another sense (Berthoz 2013). Proprioceptive sensors are fortunately at work in the vast majority of people, whether one is sprawled on a couch or adopting an awkward yoga posture. Skin sensors and vestibular regulators are active, at least at the beginning of the action, because the brain attenuates this sensation to deal with the actions to come. There are also sensors associated with muscles, tendons, ligaments and fascias that inform the brain to update its topographic map (Bordoni and Zanier 2014).

The fascias are also considered a kind of topological guide for intra-body exchanges. This concerns extra-canal fluid circulation, which is little known and yet exists since fluids circulate continuously between the blood vessels, the lymphatic channels, the nerves and the interstitial fluids. This circulation necessarily leads us to consider the presence of oriented flows, since we do not imagine that our body is a bag of stagnant water or a chaotic vortex. On the contrary, the body can only be the framework of currents, possibly guided by the network of fascias and orchestrated by the diaphragmatic breathing and the autonomous movements of the organs.

Let us explore the idea of a body crossed by fluids. When we are sitting or lying down to daydream, although apparently immobile and desynchronized, everything moves within us and in an orchestrated way between the great pulmonary, vascular, neurological, lymphatic or cerebral-spinal networks. All this fluctuates at the slightest gesture. The fluids move and exchange between tissues and organs according to the internal pulsatile movements of the body. Pressures are exerted, which can be reinforced by compressions and extensions on targeted tissues, as in yoga practices. Vascular constriction and dilation occur. Lymphatic and alimentary flows are sometimes jostled. The bladder fills up. But science is relatively silent on the circulation of fluids in non-anatomical spaces. The balancing effect of these fluids on health is a recurring proposition of "alternative" therapies. My curiosity led me to delve into the scientific literature related to Chinese and Indian medicine and to discuss with several specialists I had chosen for their solid culture in Western and experimental medicine. Here, we are at the doors of a branch of engineering that is not very well validated by experimentation according to the intangible precepts of Claude Bernard that are the foundation of Western medicine. According to these principles, the benefits of acupuncture are statistically controversial in terms of therapeutic benefit in humans, although modeling in animals demonstrates effects (Zhang et al. 2014). But viewing the body as an envelope with a structured hydraulic network running through it allows for theorizing fluid dynamics objectifying meridians. The model proposed as a result of sophisticated technical developments is that meridians are formed by true fluid circuits of "low resistance" that do not need anatomical structure to flow and be guided (Zhang et al. 2015). The future will surely better characterize these fluid and energy exchanges. They needed to be mentioned in response to the many people in therapeutic wandering who use them. An example is provided by the emergence of emotional freedom techniques (EFTs), which combine skin stimulation on acupuncture or pressure points (trigger points) with adapted gestures and psychological verbalization. The clinical interest of this technique has been documented in cases of psychological distress (Shahar 2020), whatever the mechanism.

To come back to proprioception, it is permanently required and synergistic with the other sensorialities. A good example is that of the juggler. They will look at their balls and anticipate the action of their hands (calculation of mass, inertia). They will feel the balls (haptic sensation, evaluation of viscosity and elasticity). Proprioception will help them to define the position of their arms and hands and appreciate the possible movements according to the degree of freedom of their limbs. They will hear the balls landing in their palms. The brain will process the coherence of the information and the actions to be done: managing the other hand, adjusting the eye pursuit of the gaze, tuning the speeds of the ball and the hands, positioning the hands to receive a ball, stabilizing the gaze on the right plane by the position of the neck and the shoulders, maintaining balance because of the spacing of the feet and the curve of the spine. The brain will calculate the deviations from the perfect gesture that its brain has stored during training; it will catch its mistakes and adopt a different strategy if necessary. It will call upon some sort of nervous platforms to ensure this multisensory fusion and convergence and will eventually activate what some consider to be the two brains in order to remain both logical (left hemisphere) or creative (right hemisphere) and in order to adopt a strategy that is either more compliant (following the rules) or more liberated. It is very simple, and we will start playing ball tomorrow!

6.7. Touching and stroking to promote elasticity

Nerve fibers in the skin capture information about pressure, temperature, acidity or electricity. This communicates information about gripping an object, the possibility of a scald and materials to avoid. In contrast to this risk management, touch is the vector of caressing under the hand, the softness and warmth of skin against skin or the feeling of bare feet on an abandoned beach. The sensations of skin and mucous membranes exacerbate sexual games. Touch also enables juggling with the objects, correlating a haptic command with the information of vision, hearing or proprioception, as during manual practices or juggling.

The skin is an envelope that covers the body, and as such, it could be univocal. However, there is a slight difference in potential between the upper and lower parts of the body. This has been demonstrated by physicists studying the earth's electric field, which is positive in the upper layers and negative at the earth's surface. It is in this electrical gradient that human life is played out. What started out as a simple scientific exploration gave rise to a therapeutic approach called Earthing or Grounding. The tests consisted of comparing the electrical gradient of people in contact with the earth, for example, by walking barefoot or wearing non-conductive shoes. Because of this simple protocol, the "grounded" people rebalanced the electrical gradient annihilated by wearing non-conductive shoes (e.g. leather or plastic). The collaboration of physicists with clinicians has enabled the testing of this grounding protocol on people suffering from chronic diseases. Beneficial effects have been observed on chronic pathologies, which is of particular interest to us as they involve multiple syndromes involving what might be called dys-elasticity (Chevalier et al. 2012). Clearly, a good doctor will advise caressing the bare foot in the grass or on the beach!

The alternation of hot and cold is presented as advantageous for the body mechanics and the maintenance of elastic capital. Indeed, to cool down, under the control of the hypothalamus, the body can widen the dilation of vessels, which increases the radiation of heat through the skin, together with an evaporation due to a secretion of fluids by the sebaceous glands. The respiratory rhythm accelerates, which leads to an increase in the convection of internal heat from the lungs to the outside air. The opposite is true when it is cold. We observe a constriction of the veins which limits the diffusion of blood in the body to keep the heat and avoid its radiation. Blood pressure increases, and if necessary, the brain sends a signal to the muscles to shiver, to create and release heat. The thyroid adds to this by sending a signal that accelerates cell catabolism and the associated heat production. The alternation of hot and cold is therefore like a workout for our veins and arteries (Tansey and Johnson 2015). For example, moxibustion in Chinese medicine, by applying heat to the skin, induces vasodilation and relaxation, while cryotherapy causes constriction, which can be beneficial for temporally reducing inflammation and pain.

This theme of the vital link between vascular mechanics and temperature regulation was supported by a study showing a potential alteration in elastic fiber formation. Mice mutated on a protein associated with the formation of these fibers were selected at Washington University in St. Louis. These mice did not show any particular alteration except for not producing brown adipose tissue (Craft et al. 2014). This link between elastic fibers and adipose tissue was perfectly unexpected.

Brown adipose tissue is stored in wintering animals where it generates heat and increases glucose consumption. Brown fat has recently been identified in humans, where it is associated with maintenance of normal blood glucose levels and thinness. It is found on the upper backs of infants and usually fades afterward, although some remains on the neck. The question that arises is obviously how to promote brown fat, since it has beneficial effects even for us who are not all woodchucks. This would be all the more interesting since this brown fat burns the normal fat known as white fat. It is known that the cold activates the formation of brown fat. We can start the training by finishing our shower under cold water, for example, just the areas often exposed to cold such as feet, legs, neck or head. But it will surely take a little more to get that beneficial brown fat. Fortunately, nature is good, and white fat can turn brown and beige, having acquired the fat burning and heat shielding capabilities of brown fat. Exposure to cold still appears to be necessary for this browning (Okla et al. 2017). Another avenue currently being explored is to promote the gut microbiota. While waiting to better define how to promote the formation of beige fat, we can only suggest spending a few vacations in a cool place, far from the tropics, to benefit from the temperature of water and air that will stimulate all elastic capacities and the synthesis of beige fat. This will be good for the body and will reduce the tourist pressure of warmer water areas. At the extreme, we can cite Wim Hof, the ice man, whose positive effects of cold training on his body are carefully monitored by scientists.

6.8. Hearing and feeling in order to enjoy the song and movement

At the soft end of the scale, speaking and singing are good examples of simple and accessible practices that, although not specifically therapeutic, contribute to the maintenance of elastic capital. In this book, I have already described what happens when we sing or dance, in order to make us understand how much an action as simple as singing, right or wrong, and dancing, elegant or clumsy, has implications in the body. When well done, singing mobilizes the breath, from the feet to the head, passing through the perineum, the diaphragm, the lungs, the vocal cords and the nasal resonators.

And reciprocally, hearing benefits from singing since it activates all the mechanical parameters essential to hearing.

The voice, spoken or sung, can stimulate the transmission of sound from the vibration of the eardrum at the entrance of the external ear to the elastic blade supporting the tissues of the cochlea and its sensory cells. We recommend soft words and songs with the tip of the lips to avoid damaging the elastic and collagenous fibers of the vocal cords. Sound feedback can be interesting so that the brain is aware of the sound emitted and optimizes the tension of the vocal cords because of the tension of the cartilages of the larynx. This facilitates the exercises to fight against age-related hearing loss.

Hearing calibrates the gestures of dance. It offers its harmonies, rhythm and intensity. Music creates a strong social link and makes the body go beyond its limits. The body's energy networks are put to work. The energy is networked because of an orchestrated deployment of the production of ATP. Breathing is deep and helps regulate the body's acid-base balance. Cognitive stimulation is important and reinforces cerebral plasticity. All these positive effects are undeniable in pathological situations such as Parkinson's disease (Chastan and Decker 2019).

Olfaction is the second system of the human body in terms of genetic endowment, after immunology. This shows its importance in the mechanisms of recognition, socialization, orientation and defense mechanisms (Pierron et al. 2020). This also means its role in obtaining a reward and the stimulation of the emotion. Perfumers, florists and cooks know something about it. This is one of the reasons that led me to introduce in this book the flavors of aromatic and therapeutic plants, spices and vegetables to optimize a reasonable diet. It is also a reason that pushes therapists and rehabilitators to be interested in the cognitive aspects of olfaction. Some smells have a positive emotional connotation, i.e. they can increase motivation and promote learning in children and adolescents with autistic disorders or during rehabilitation sessions after a physical trauma (Luisier et al. 2015).

Hearing, olfaction and gustation contribute to the elasticity of the body according to each person's resources and motivations. Several applications can be implemented such as improving the learning of writing with a sonic pencil, or improving postural gestures and proprioceptive and social perception of people suffering from gesture disorders or autistic disorders.

Diction, breathing in a natural and oxygenated environment, the search for smells, the discovery of culinary specialties, the cultivation in a garden or in a pot are all activities rich in emotions. They should be used without moderation! A somewhat neglected aspect of tasting concerns its mechanics. While the taste buds inform us of multiple flavors, the tongue and chewing inform us about the texture of food by mobilizing touch and proprioception. The jaw's ability to move is made possible by a complex joint. Associated with its muscles and ligaments, it allows swallowing, chewing, phonation and yawning. Suffice it to say, it is forced to contribute a lot and that it struggles going the distance, especially when elastic capital is lost. So, we should take a good look at our jaws, teeth and tongue with as much interest as feet; without one we would not walk, and without the other we would not taste much.

The terms "anosmia" (loss of smell) and "ageusia" (loss of taste) were relatively unheralded until the recent insertion into our world of a virus affecting these sensitivities. Many people have discovered the virus in their flesh because of the loss of these senses. The virus has shed a sudden light on situations that are very disabling because they deprive people of a very important element of defense and sociability. They must solicit other sensory resources, such as vision or touch, to support their assessment of the environment. This is obviously what people with sensory loss do, when the resolution of intimate or "extimate" problems requires the mobilization of other tools and other coping strategies. Apart from these considerations, updating the notion of anosmia is also an opportunity to share some understanding with people with olfactory and gustatory hyper-sensitivity. Their lives can become a living hell when hygiene and cleanliness are not respected around them. This last comment will serve as a transition to the moral, social and societal issues, which will be discussed in the next chapter.

6.9. Conclusion: let us adopt a loop of "pro-elastic" postures

Each person can make their own choices when faced with the profusion of possibilities to stimulate their elastic capital in order to maintain, enrich or protect it, depending on their experience. There is no definitive answer to the question of adopting the right postures, except to move without doing more harm than good. Personally, I have developed my own series of stretching and pro-elastic strengthening exercises, i.e., exercises that "do good" to my elastic and collagen fibers. My protocol incorporates a mix of postures gleaned from gymnastics, Pilates, yoga, fasciatherapy and physical therapy. Nothing complicated, in fact, but with guiding principles dictated by the loops of compression/extension and vasodilation/vasoconstriction that move my musculoskeletal system and my organs. We often neglect our organs, yet they are often the ones that need elasticity the most. We can base ourselves on practices that incorporate notions of biomechanics and physiology such as those described for example in yoga postures to "prevent and heal" (Campagnac-Morette 2021). On the sensory level, an adapted musical rhythm accompanies me while a mirror allows me to check my posture and let my own mirror neurons have fun in my inverted image. I also use elastic bands, which allow me to gently implement the very rich notion of leverage. Finally, I recommend walking to everyone, especially Nordic walking, which is a very stimulating and complete practice for the elastic system, with the advantage of being practiced outdoors, of being inexpensive and of facilitating sociability. It is also mentioned in the scientific literature as a favorable practice in many rehabilitation situations (Roy et al. 2020).

We must talk about this notion of leverage because it conditions the success of many efforts and preserves our energy by limiting breakage. There are three types of lever arms: the so-called "Roberval" two-pan balance, the wheelbarrow and the tweezers. The lever arm of the balance has a support in the middle. It corresponds to our head carried by the Atlas vertebrae. The support is in the middle (on the Atlas), the weight carries forward and the back muscles keep the head straight so that it does not lean forward. If you are already bent over, the neck will get tired more so we stand up straight with our head up during the exercises! With the second lever arm, that of the wheelbarrow, the support is at the end on the wheel of the wheelbarrow, the weight is in the middle and the effort is on the arms. The longer the arm, the easier it will be. This is the situation for the feet; the support is at the end of the foot on the first metatarsals, the weight is that of the body supported on the ankle and the force and the lifting force is at the level of the leg muscles. This explains how important the support on the first metatarsals is, which is something I keep telling my teammates who practice Nordic walking or cycling. The third lever arm is that of the tweezers, when the force is applied in the middle between the support (the junction of the arms of the tweezers) and the tip on the hair to be removed. This lever arm is at work in the forearm when the anchor point is the elbow, the force is the muscle and the hand holds the weight, hence the validation of the "long arm". These lever arms must be taken into account in all efforts, especially in relation to the spine.

Here we are, ready to look at the elasticity of the mind and the research and implementation of effective leverage arms to address and optimize this elasticity through an approach at the personal, social and societal levels.

7

The Fourth Challenge for the Elastic Human: Adopting the Spirit of the Laws of Elasticity

Figure 7.1. *Elastic capital and otherness. The intention of the sixth painting is to illustrate the notion of otherness. Stereotypes exist that determine our behaviors, questions await answers, caregivers are involved, scientists theorize and develop research, realities intersect (Dussart et al. 2019). Characters from the painter Witold Wojtkiewicz were called upon as their expression is so strong, especially in his painting* Christ and children

For a color version of all the figures in this chapter, see www.iste.co.uk/sommer/elasticity.zip.

7.1. Introduction

The description of the body's abilities in this chapter exposes the remarkable aspect of our elastic capital. Reed Richards, the hyper-elastic comic book hero capable of stretching beyond the seas, is not the only fantastic one; we all are, when all goes well. From basic natural materials, infinitely richer and more complex than described in this book, the body fulfils its vital functions of movement, sensoriality, communication and reproduction. The fusional hybridization between the body and the brain is a masterpiece of efficiency. The vast majority of humans are born with a consistent elastic capital and do not ask themselves any questions. So much the better and we have Mother Nature to thank for having selected such a well-constituted vehicle. Everything could be at its best for the elastic human. However, not everything happens exactly like that; otherwise, there would be no need to write or read this book. In order to reflect on what causes problems, we must venture into the lands of the human sciences while respecting the biological perimeter of the elastic human. We have previously outlined the achievements and challenges of biology and technology to study, repair and even compensate for a deficit of elastic capital. We have invoked the adequate food necessities to build and maintain this capital. We have linked sensorialities and body mobilities to maintain and optimize our body. We have (quickly) mentioned some controls of the central and peripheral nervous system. It is now time to position the living being's elasticity in the social and societal context, where it plays a role that is far from being trivial as we have seen repeatedly in this book.

Indeed, the deregulation of vascular pressure is often part of the same societal chain linking advertising for hyper-sweetened products to deforestation for the feeding of livestock while using intensive farming and sedentarization following the development of the use of digital tools. The selection of proliferating or resistant pathogens is at the end of a chain of events that includes the inappropriate use of antibiotics for industrial breeding, the sale of wild animals on local markets and the globalization of transport. Many chronic diseases are in some way linked to industries that are disconnected from their biological responsibilities. In the extreme, one could even link the fate of our elastic capital to behaviors and choices dictated by economics and politics. The extension of the concept of human elasticity to society is therefore not a trivial matter. In order to focus this debate and to remain within the limits of this book, we will frame our discussion with a few laws from physics, mechanics, chemistry and biology that are already used in the human, social and societal sciences.

As a reminder, we have already applied laws of physics (elasticity, compliance, fluid mechanics), chemistry (cross-linking, gas diffusion) and genetics (genetic determinism mixed with epigenetics and the influence of the environment) in the first three challenges of this book (managing the body's elastic capital, feeding it, optimizing it). It was already a real challenge as the scales of analysis differed. We also encountered several points of contact between these laws and society, for example, by talking about the effect of pollution, unreasonable diet and screen-based sedentariness on the loss of our elastic capital. This requires us to explore these areas. We are indeed destroying our capital through disproportionate or repetitive efforts often linked to work, to the pace of life and to the speed of our activities, through inept diet and behavior, through the orchestration of an inequitable society or more globally through the war of attrition that humanity is waging against its host land. To remedy these situations, technical, economic and political models are being studied that will invoke, for example, a frugal engineering or an equitable sociology.

The opposition between nature and an elastic human could appear somewhat opportunistic in the current context of global warming and mass extinction undoubtedly caused by human activities. It would be out of place if it did not have the ambition to lead a reflection on the value of laws defined in the field of the so-called "hard" sciences but used in social practices, as the late philosopher Jean Gayon, interviewed by Victor Petit, perfectly described in his book *Knowledge of Life Today* (Gayon and Petit 2019). I quote him to thank him for opening one of the CNRS colloquia on the "augmented human" that I co-hosted. To give an example of the fragility of these laws, Jean Gayon evoked the interpretation of Darwin's initial work about the supposed law of the jungle: may the fittest win! This misinterpreted concept, however, reflects a very partial vision because it excludes the notion of mutual aid that Charles Darwin developed in relation to humanity at the end of his career. This law of mutual aid, which offers an equally powerful selective advantage, is strongly fueled by epigenetics, that is to say by the sociological and natural environment. It opposes the selective advantages of mutual aid to the supposed benefits of competition. In reality, the initial bio-sociological studies (by Darwin and Pierre Kropotkin in particular) as well as more recent ones demonstrate how the interests of the individual and the group can be combined in a flexible and reciprocal way to build a selective advantage (Kropotkin 2009; Servigne and Chapelle 2019). Evolutions are thus possible through the reinforcement of the empathy of individuals because of the education of the group and of each individual within the group. So here we are, with a self-help law to add to our panoply of effective tools.

The law of mutual aid is already present in everyday language, as are several laws of mechanics that will serve as semantic answers to bring the elasticity of the living into the world of the human and social sciences. We will start with cross-linking, which has been positioned under the umbrella of the notion of network, strongly challenged by a thirst for knowledge and pseudo-absolute truths that are expressed through unstructured digital tools. We will continue with resilience, which has become the star of psychology under the enlightened baton of Boris Cyrulnik. Compliance resonates with morality in the specifications of companies and the clinical monitoring of therapeutic treatments. As for elasticity, which is ubiquitous in this book, it is also used as a synonym for the price flexibility that characterized the law of supply and demand, which has seen its usual practice evolve from a hand-to-hand exchange between elastic operators to a computerized trading system in which all notions of naturalness have disappeared.

7.2. Cross-linking and knowledge

Cross-linking will be the first key word chosen. It comes from physics and chemistry. Cross-linking concerns the formation of links to secure and structure an object through a network of interactions. It is through this type of action that body fibers are reinforced for better or worse; better because elastic fibers will not be functional without cross-linking, worse because a deposit of fibrous material that is too cross-linked will be toxic, as in fibrotic lungs leading to respiratory failure. The word cross-linking will be used as a network with a view to consolidating and sharing personal and collective knowledge.

We all know intuitively the rights and wrongs of social networks, whether they are physical or digital. The first computer networks were developed for military purposes and then for exchange between academics. Interactive platforms then came into play, making available knowledge that is accumulating at an exponential rate. Exchanges between scientists have become more fluid, while patients have become better informed, some even earning the title of experts. As every new quality gives rise to ugly flaws, the beneficial increase of information sharing through social networks has been associated with a cohort of pseudo-science and fake news, as well as with an outpouring of hate and harassment, not to mention deliberate attacks on freedom by governments and pressure groups. These outrages flourish in times of constraint, pandemics or environmental crisis. On the other hand, an abyss has been dug for the digital castaways that do not have the means or the skills to access the tools that have become the new standards for administrations. However, we will take for granted the satisfaction of having strong knowledge networks, fully aware of the risks associated with their inadequate use.

7.2.1. *Knowledge and the expert patient*

There are several ways to cross-link and assemble parts of a material which do not necessarily lead to the same result, as we have seen previously. Similarly, there are many ways to share or subjugate knowledge. Once again, pandemics are instructive in this respect, as many supposed truths have become mired in sterile debates while waiting for confirmation or denial.

Knowledge about biological systems is part of this constantly evolving knowledge, all the more so when it appears very fragmented, as is the case for the notion of body elasticity. The knowledge is first of all scientific and technical, and it is broken down within areas of specialists who interact relatively little. However, in this fragmented form, it is not available to the people concerned, either because of their individual difficulties or their curiosity and commitment, which constitutes the first networks of expert patients. This makes it difficult to get an overall picture. The network dedicated to cutis laxa was initially of this order, as are many other networks initiated by a destitute family faced with chronic or acute suffering with no therapeutic solution (Boucand 2018). The complexity of the knowledge required to develop hypotheses about the long-term consequences of COVID-19 is part of this difficulty. There may be a limit to medical knowledge, uncertainty in diagnoses, and hesitancy on the part of those affected when faced with the appropriate remediation to adopt (as is the case with vaccination).

Knowledge can be intuitive and relative when it corresponds to each person's feelings. The pain of one person is not the same as the pain of another, despite the fact that the causes are sometimes similar. Pain can be a hell of a thing that is difficult to describe and talk about. This is all the more true for shifting disorders, different from one day to the next, depending on fatigue, air, rain or lack of sleep. The evocation of psychosomatic causes by experts who are sometimes powerless testifies to this powerlessness. Fortunately, knowledge is organized and shared through the experiences of others. It is in this space where unknowns abound that intuitive knowledge is inscribed, often amplified by an absence of therapeutic solutions. This is the case for many people I have met in various rare disease circles. Their network necessarily expands, similar to lungs sucking in the air of life.

Knowledge is built in an "experiential" way when each person identifies deviations from the norm that are difficult to determine or apprehend, because they are latent, chronic, sometimes evolving, and often painfully diffuse. The recent evolution of social networks has facilitated the possibilities and organization of sharing. These discussion spaces allow an unprecedented enrichment of personal intuitive experiences, taught by sharing. This also feeds the preventive aspect which is the poor relation of medical and intuitive knowledge. Knowledge takes on an empathetic and therefore resolutely ethical dimension. The individual good becomes a shared common good.

Communication between individuals and associations that have become experts in a health field by necessity is an excellent way to concatenate isolated cases. It is also a trap when communication remains focused on particular cases, which weakens the political scope of the whole. Fortunately, associations know how to gather under the banner of alliances, such as that of rare diseases at the European level. This is a real societal change, when the representative user of an association becomes a legal person who is no longer the "notorious incompetent" that they remained until relatively recent times (Boucand 2018). They become capable of verbalizing ailments and better understanding where their place is in the value chain within the health system.

Let us go back to care or health pathways. The user's verbalization, the more or less specialized diagnoses, the research, the medical treatment, the therapeutic follow-up, the support after the care, the possible side effects and undesirable effects, the costs of care and the relationship to knowledge in multiple networks build a chain of actions. The user, integrated into their network, must be able to intervene in this dynamic, and this is one of the criteria for the effectiveness of the care pathway. This is all the more true because of the digitization of user data. Other players are also involved, such as health companies that manage patient data, insurers, and sometimes employers, as well as political leaders and even health security agency directors. The science of "data" is shaping a participative and integrative context that enormously enriches "human-scale" medicine or participates in the well-being of everyone. The Silver Economy, which targets the well-being of people aged 50+ and too often suffering from chronic diseases, is part of this value chain at every level. We can therefore understand the importance of integrating the notion of declining elastic capital into this chain of actions.

The integrative process of empowerment based on collective intelligence and integration in a chain of values testifies to a very elastic and adaptive spirit of the actors of care, users and caregivers (Dussart et al. 2019). Certainly, the inclusion of users in the reflections that lead to public decision-making can induce an injunction to legislators to adopt more binding standards. But much remains to be done in view of the situation of people with disabilities and their caregivers. However, standards can and must change, under pressure from groups, whistleblowers and the evolution of knowledge. Once again, let us note how pandemics have gone down in history for having imposed adjustments in the sharing of knowledge and the evolution of public health policies. The day-to-day diagnosis, the protective measures, the economic prioritization, the marking of infectiousness, the therapeutic development and simply the taking into account of the fragility of life are so many elements that have and will lead to repercussions on all the elastic human's capabilities.

7.2.2. *Knowledge and cyberchondria*

Networks exchange their skills but also their incompetences. As was the case for AIDS, the worldwide networking of media coverage of a pandemic illustrates the dynamics or the errors of the debates between these three types of knowledge: medical, intuitive and experiential. This communication sometimes demonstrates by the absurd the limits of all self-proclaimed experts when faced with the complexities of biology. History will tell toward which truth they will converge. But it is certain that everyone can search on networks for what concerns them personally. And everyone will find it, because the networks are above all a very effective echo chamber for our hypochondriac tendencies, which we will find all the more relevant as our knowledge is weak and our questions strong, as is the case for the ailments associated with loss of elasticity.

Everyone will find the right disease on the Internet, in all the jumble of information and peremptory opinions available (Bronner 2021). The Internet is the bread and butter of the new cyberchondriacs, who strive to find knowledge there that echoes their health concerns. This is, among other things, a consequence of the "cocktail effect", by analogy with those situations when our sensory system individualizes the voice of a distant speaker among the hubbub of the party. This attention will be all the more easily acquired if the information gathered irritates us or arouses our curiosity. It is a cognitive skill, when discerning scrambled information could save our life. This is another feature offered by evolution that will save us from an unwanted encounter at a cocktail party or allow us to target the person we absolutely had to see. The assemblers of network platforms have understood this, and their algorithms will pleasantly tickle our reward neurons. The consequence is a cognitive explosion that is difficult to resist, especially when certain knowledge becomes dogma. The stakes are high and must not be left in the hands of commercial or autocratic interests. This being said, "chance (*to acquire the right knowledge*) favors the prepared mind", to revisit an expression of Louis Pasteur about the discoveries of scientific research, thus reminds us of the importance of sharing duly documented knowledge.

To conclude and return to the analogy of the elasticity of the body, the social cross-linking of the world through its digitalized networking can have beneficial effects as long as it is organized and not anarchic. It is because of a controlled and non-anarchic cross-linking that the elastic capital is optimal. It is through the sharing of coherent knowledge, properly fed by quality scientific research, that the world can progress. But it is through anarchic, non-referenced and even voluntarily subjective interventions that social networks have become the sounding boards of a world losing its bearings. However, these detuned words had few places to express themselves before. It is therefore necessary to continue this sharing, but not to transform it into a superpower capable of turning the world into a cognitive tyranny. Vigilance and exigency are required, at the global, national and local levels. This book presenting the elastic system of the body is part of this approach to sharing knowledge.

7.3. Resilience and the mechanics of the world

7.3.1. *Resilience and the elastic human*

Resilience was originally a term used to describe the ability of a material to resist pressure and absorb a shock by deforming. It is now used to describe a predisposition of mind and character for people who positively continue their life journey after experiencing trauma. It is a situation shared by all families including people exposed to situations of disability or loss of autonomy. Society itself is looking for resilience after the shock of a pandemic, such as that caused by the coronavirus family (Emmanuelli and Cyrulnik 2021).

The human body has incredible assets, only one of which is addressed in this book. We could make the same observation with our other bodily capitals whose diminution, loss or lack requires a process of resilience. The identification of a biological capital often takes shape in front of the realization of its excesses or of what it is lacking. A feeling of unheard-of violence falls on children and their parents when an anomaly appears. Sometimes, as was the case for the little girl we have been following since the beginning of this story, some signs suggest a deviation from the norm. Then, the anguish settles and will be amplified if the dysfunction, the disability or the disease is confirmed. Some will correctly experience this strange situation, with its liminal zones with fuzzy contours and zones of intensification of differences that will have to be named in order to adapt well (Boucand 2018). Others will experience painful scenarios when genetic defects, disease or trauma impose their new norms. The loss of capabilities will threaten these unlucky people. Among these will be the many people who will live with the consequences of chronic diseases like COVID-19 and the whole panoply of dysfunctions that affect the biomechanics of our bodies that we have already presented.

Any crisis provokes a reaction and sometimes even a shock wave, which may have an internal cause, when the body and mind unravel, or external when the environment exerts an inexorable pressure. We will then speak of shock and pressure, but also of resilience, adaptation, uncertainty, instability or adjustment, all terms that can refer to a mechanistic or psychological reflection. I leave it to the neuropsychiatrists and psychologists to enlighten us with their expertise. However, I can testify to the extraordinary resilience shown by the many people I have met who suffer from a moderate or severe limitation of their elastic and sensorimotor capital, whether innate or acquired. And yet, this resilience is all the more difficult to establish since the notion of elasticity of the living being in all the intricacies developed during this book is little or not known.

The current era has seen the development of a great interest in the search for well-being, personal or collective fulfilment and the power to act. This was not the case before, and it is not the case everywhere in the world. However, it has become a dominant discourse that can be very interesting for the users because this dynamic identifies psychological, qualitative and quantitative criteria that build a real system of well-being or health. We can therefore validate the "well-being" factor because of psychological markers as we do with physiological markers. This brings a validation of practices that I have been able to introduce in this book. It is a question of considering all the components of the system and of comparing the corresponding therapy or support solutions. Finally, the suffering person or the group considered is no longer observed through the small end of the physiological spyglass but through a global approach for what it is, i.e., a complex, intelligent and performative entity. "The solution-focused systems approach" responds to the demand for a global vision objectified toward concrete improvement (Dussart et al. 2019). It must be reiterated how utopian this holistic vision would be if it were not facilitated by the growth and sharing of knowledge.

7.3.2. Resilience and society

The mechanisms of resilience do not only concern the individual; the whole society can be involved: school, family, work, meetings and leisure. The psychological management of the invisible difference is particularly difficult. I remember talking to a person affected by an invisible illness (a hyperlax Ehlers-Danlos syndrome) but who continued working through sheer will and the help of compression bandages that gave her a fragile stability. Her day must have been exhausting but nothing showed. What a constant control she required! The visible difference is also very complicated, at school, when integration is difficult, or during adolescence, when stereotypes become very prevalent and result in mockery, isolation or harassment. Several syndromes concerning body mechanics evoke these pejorative representations which have led to the undesirable naming of stereotypes of dwarfs, giants or elves. But there is no exclusion *a priori* if the support is well organized, which is not always the case and needs to be greatly improved by taking into account otherness.

Caregivers clearly have generous and imaginative traits to protect by positive stereotypes the fragilities linked to dysfunctions, differences or even a lack of autonomy. They have created associations, made them known, created supra-themes and acquired knowledge requiring years of expertise. One of their most endearing characteristics is their sense of giving and sharing, all the more so if their understanding opens up to all situations of disability or bodily insufficiency. This vision enables going beyond the corporate status and increases the possibilities of resilience in all its sociological dimensions. Mutual aid is an important factor of resilience.

Humanity has been able to add through its technology an unheard of factor of amplification of our individual mechanics, which has enabled sending men to the moon or creating personalized bionic hybrids. The long list of scientific discoveries has made it possible to extend the human body space almost infinitely. The elastic human is no longer the one who used to ride 10 leagues a day, at most. These 40 km are now multiplied by tens by driving a car, or by hundreds by flying by plane. This change of scale changes the scope of human life. The space-time of humanity has contracted or dilated. As a corollary, by dint of technologies that push our limits and multiply the scope of our actions, quickly and comfortably, our societies have forgotten the magical nature of our body envelope. These are the points of reference that we must remember in order to resist the dynamics of a technological dehumanization. It is all the more necessary when the necessity of a societal resilience is imposed under the constraint of an infectious pandemic (Emmanuelli, Cyrulnik 2021). The effects of this constraint are already apparent, with the consequences of the deleterious actions of a hyper-mechanized human being in their incessant wars for the control of the earth's resources, be they mineral, vegetable or animal.

I want to take as an allegorical example a fiction by Jacques Spitz, who is a French author from the middle of the 20th century. Known by a limited circle of science fiction lovers, I chose to quote this author for his 1938 novel, *L'Homme élastique*, which justifies the analysis of his text in this writing. His hero is a scientist who has discovered the possibility of reducing or expanding the molecular scale of life. Concretely, his technology reduces or enlarges men at will. Their elasticity in this context is perceived as a change in size and a return to the original form is always possible. But the most interesting part is in the exploration of the consequences of a loss of reference marks associated with what one could qualify as the size of the body, through the modification of an essential physical capacity which is the size. It is firstly the French army that seizes the project to reduce soldiers in order to infiltrate the enemy armies incognito. The enemy was defeated because of this process (one can see there an anticipation of the effectiveness of espionage). The rest of the fiction exposes the author's point of view on the intentions of certain countries. In France, it is a cacophony and everyone does as they please. The tendency is to reinforce and increase the body in Germany. England refuses the process on the basis of its principle of habeas corpus and the United States opts for a specialization that generates more productive efficiency. Finally, in the story, humans will have to adapt to the myths of the Nations and those who refuse the change of scale will be incarcerated in a camp. The scientist will die alone and thrown back into one of these camps.

This parable of the "elastic human" prefigures, in a way, all the errors of the technological expansion of human elasticity. In reality, it does not prefigure it; it already figures it. For the unlimited increase in the scope of our actions is already leading humanity to a sixth extinction of the earth's species and a climatic upheaval that is only just beginning to appear right in front of our already much encumbered nose. The health of humans and especially the elasticity of the living are chronically affected by the unreasonable procedures of agribusiness and chemistry. The hypermobility provided by the motorization of the vehicles and, on the contrary, the sedentary lives caused by the omnipresence of digital tools are taking their toll. Behavioral injunctions facilitated by technology are surreptitiously introduced into the heart of our world. The list is long and growing. Jacques Spitz's fiction is overtaken by reality. The shock for nature is harsh, and harsh will be the environmental upheavals generated by this expansion of the elastic human. The adaptation of societies to this shock is on the horizon. We are starting to talk about financial, entrepreneurial and Anthropocene resilience. Technical or political solutions are being discussed, but the holistic tools of a societal resilience to be orchestrated are not yet known.

The adaptations outlined during a global health crisis foreshadow the many ways in which social and societal resilience can be sought. But we should not be mistaken in believing that an endless expansion without major changes will continue without deleterious consequences. For one of the laws of population genetics considers that no species is in equilibrium. In concrete terms, this means that the major species always reach a point of inflection in their growth followed by a decline when they alter their environment to the point of making it unliveable. Other species then take the lead. This evokes a kind of selection through the collective behavior of viruses and other pathogens that proliferate to the detriment of a humanity that stubbornly destroys the lightning rods that protect its environment. To avoid this, we must anticipate the consequences of our actions and conduct more respectful strategies. This requires a permanent examination of our social and societal norms, exploring in particular the capacities selected by evolution, whether they are cognitive or behavioral. We must once again emphasize the power and limits of human physical and mechanical capacities, which flourish best when a safe and egalitarian societal climate is felt, facilitating trust and mutual aid.

7.4. Compliance, ethics, law and elastic capital

7.4.1. *Compliance*

Compliance is the stretching capacity of a material. This physical parameter defines distensibility and flexibility. It is a concept that we have seen with the example of laddered pantyhose whose capacity for extension is increased, without keeping the original elasticity. The treatment of the pantyhose with nail polish avoids the complete unwinding of the mesh, without preventing the formation of unsightly holes. The body does the same by bringing new material to the sites of pulmonary alveoli destruction. But if the stimulation is too strong, fibrosis and respiratory insufficiency will set in.

Compliance is a term used in many fields other than the mechanics of materials. It is used to qualify the behavior of patients, depending on whether or not they follow medical prescriptions to the letter. We speak of a compliant or non-compliant patient. Compliance is also used to evaluate the degree of respect of rules in companies. It is a question of compliance with the law and the values of the system. The crucial question is who defines these values, how they are determined and how they are assessed.

In the chapter on food and food inputs, we have discussed at length the toxic effect on the elastic capital of compounds that are available on the shelves of every shop in the world. The examples cited are sugars, the *ad nauseam* exploitation of animal and plant resources, additives (such as gluten and food sweeteners), alcohol, cigarettes and drugs, to which we could add packaging and the conditions of production and transportation of materials. This list is part of a catalog that calls for the evolution of criteria to ensure a level of compliance that would not be detrimental to the elastic human being.

7.4.2. *The elastic system and the technical system*

The key idea of the existence of "systems" emerges from the compliance examples mentioned above. Pantyhose is a structured polymer whose cross-link becomes more deformable and less elastic if it is filled with flakes of skin and secretions. Its primary function of harmoniously following the curve of a leg is thus modified by the evolution of the physical organization of its structure. By analogy, the non-compliant patient defines their behavior in relation to the understanding of their needs, which are structured by their education, quality of life in a specific society, position in the medical-social chain, trust in what clinicians and pharmacists are telling them and acceptance of societal injunctions. These are the structural mechanisms of thought that are at work in the social acceptance of a vaccination, to take a recurrent example that is highly topical in times of a pandemic. At the same time, states and companies establish their procedures according to reasoning that combines economic interests and governmental or legal norms. Compliance in these examples is thus defined by forces that structure social and societal systems, the analysis of which is beyond the scope of this book. But if we push the analogy with the mechanics of materials somewhat, we can consider that if there are forces, we can orient them by applying an adequate lever arm.

Most systems identify structures that actually react to forces. For example, the systems mentioned above have in common that they associate a human entity (the patient, client, user, citizen), a technique (nail polish, drug, a device or a service) and a communication (the annoyed interjection, a prescription, the advertising of the product). This triptych is well known to philosophers, but I would like to refer to Jacques Ellul, whose writings I have already hijacked to dare to introduce the horsemen of the apocalypse in this essay. Ellul, a professor of the history of law in Bordeaux, worked on "technique or the challenge of the century" (Ellul 1990). To make it (very) short, his analysis considers that the multiplicity of specific techniques builds an autonomous "technical system", self-reproducing, growing and without finality (but with specific goals). In the beginning, humanity developed tools to make life easier for itself in a need for survival and power. Since then, the race to inventiveness has offered innumerable facilities to humanity, which would be stupid to deny, but at the same time, humanity has developed, according to Jacques Ellul, a technicality that constrains freedom by becoming indispensable. It would thus be a question of better reflecting on the finalities of technology, in order to be able to frame it and associate it with criteria more compatible with the biological fundamentals of humanity and of the living, of which the elasticity with its capital that is inexorably reduced is a part. With this concept in mind, we can envisage multiple solutions targeting the "technical system" to optimize the living conditions of the elastic human, which, in essence, must respect humanity's environment. Moreover, the solutions appear easily "implicitly" in the statement of the toxic effects of unreasoned food, of pollution, of the extinction of biodiversity and the digital sedentary life.

In reality, the conclusions drawn by Jacques Ellul and other authors were rather disenchanted, given the lack of concrete mobilization to give finalities to a growth defined by productivist programs, despite the recommendations of the Treaty of Rome and the Meadows report. This disenchantment persists today because the current political or economic offer has obviously not implemented their systemic recommendations. It is because the matter is not so easy. To the credit of the participants in the many well-meaning meetings such as the Conferences of the Parties (COP) for climate change, no recent society based on technological development has achieved environmental balance. However, the concept of an elastic human perfectly aware of the strength, characteristics and limits of their body can bring a tangible parameter to the creation of a balance between humanity, nature and technology.

7.4.3. *Mutual aid and the elastic system*

The notions of mutual aid, compliance and mastered technique lead to introducing the rich notion of conviviality. This concept is perfectly suited to this book, which has already introduced many notions of behavior (Chapters 5 and 6). Conviviality is a word derived from the Latin *convivium*, which describes a pleasant meal shared together. The origin of this usage is frequently attributed to Jean-Anthelme Brillat-Savarin's treatise *The Physiology of Taste* (Brillat-Savarin 1981). In reality, this concept could also be attributed to Dante Aligheri, who, in his book *Il convivio*, which was written in a dialect that was the origin of modern Italian, proposes a sharing of knowledge and wisdom that is supposed to be accessible to all (although of an astonishing scientific modernity and of a certain complexity; Tenze 2021). The concept of conviviality therefore concerns a sharing where mutual aid is often appropriate and friendly. Here, we appear with the notions dear to the elastic human of sensory pleasure, movement and shared action. They add an ethical concept to the mastery of structures and tools by the elastic human, which may seem all the more unattainable.

The notion of conviviality has been politically stated by Ivan Illich (Illich 1973). The thought of this philosopher still nourishes the political ecology. The idea of his optimal society, qualified as convivial, is to give an ethical value to the tool, defined (or not) as "convivial" according to its accessibility and the distance it maintains with the human who uses it. Ivan Illich thus raises the question of the finality of a technique that would respect the finalities of humanity, which we will have no difficulty in associating with the elastic human in this book. It also introduces the difference between palliative, short-term solutions and long-term systemic solutions. For example, the abandonment of fossil energy by a non-polluting source is a long-term systemic necessity and its replacement by sources such as electricity appears to be an adequate palliative possibility. But in this case, the consideration a priori of a compliant injunction would first and foremost require mastering several elements of the production chain, including resources, waste and risks, before any mass use.

To illustrate the development of a tool respecting the specifications of user-friendliness, we can take the example of the control of the stump of an amputated arm by the reactivation of the phantom limb in order to drive a prosthesis. The pressure point is the stump, and the two supports are the sensory brain area and the phantom of the missing hand. The user-friendliness is strong because the efficiency of the prosthesis requires the full participation of its wearer (Touillet et al. 2018). We can find in this example the notion of the lever arm, and more precisely that of the tweezers model (see Chapter 6). Usability can also be strong when the use of exogenous robots can replace a limb after amputation. This transfer to exogenous technologies suggests the lever arm of the wheelbarrow instead. Another user-friendly alternative is to promote assistance to the disabled person by facilitating societal mutual aid. In other words, to remain within the framework of this book, it will be a matter of using the lever arm of the scale to support the costs of mutual aid and to remove them from the notion of price elasticity between supply and demand. Society can thus express a constraint that will facilitate the life of people who are struggling or disabled. The constraint in this case is to apply a force that could be described as shear, by analogy with the shear forces that act in the human body, such as those induced by the flow of blood on the walls of blood vessels.

The convergence of the notions of conviviality, mutual aid and compliance draw from two encodings of the elastic human, the individual taste for justice and equity on the one hand, and the search for reward and efficiency through sharing or reciprocity on the other. These encodings are powerful, much stronger than the notions of struggle for life wrongly lent to Darwinism (Tort 2015). In reality, collaboration between different species and within the same animal species is pervasive. The mutual aid between the strong and the weak are not illusions and examples abound in the Living (Servigne and Chapelle 2019). The association of the bacterial flora with our intestinal amalgamation perfectly illustrates this reciprocal interest, when the organism is balanced and that the partnership is not upset by a toxic competition, as at the time of the pandemics.

7.4.4. *Culture and the elastic human*

Culture enriches without question the triad of humanity/nature/technique. The iconographic evocation of the human body is omnipresent. It is moreover by using a painting by Goya representing a body in distress that I chose to introduce this book. Painting has made a profit with the evocation of the body, afflicted in Francisco Goya's *Disasters of War* (Figure I.1), iconic with Jean Lurçat's *Song of the World (Chant du Monde)* (Figure 1.1), evanescent with Odilon Redon (Figure 2.1), hallucinatory with Francis Bacon and parodic with Pieter Bruegel the Elder (Figure 3.1), descriptive with Georges Chicotot and malicious with Georges de la Tour (Figure 4.1), seraphic with Paul Sérusier (Figure 7.2) or aggressive with Witold Wojtkiewicz (Figure 7.1), as well as modeled with Vieira de la Silva (Figure 6.1). It could have been proportionate with Leonardo da Vinci's *Vitruvian Man*, elegant with Dürer, tormented with Schiele, decomposed with Picasso, ritualized with Gauguin or elusive with Raoul Dufy. Writing, sculpture (with Alberto Giacometti's *Walking Man* cited in Chapter 6), song, dance, cinema and the circus arts, among others, are not left out to represent the elastic human in all its states. The fashion dresses these states in a somewhat hypertelic approach when the accent is put on the sublimation of the body and not on the practicality of use. Culture creates the link between nature and humanity as well as between technique and humanity. In this, it is fully in line with the processes of cross-linking, resilience and compliance which participate in the formation of laws, their evolution and their acceptance by the elastic human.

We can consider culture as a system that questions first the eternal question of the relationship of humanity with nature and morality. Culture feeds the notion of humanity, for the good and for the evil of it. But humanity has known and knows in its flesh slavery, alienation of men and especially recurrently that of women, racial supremacy and inferiority of birth. The tortures and malevolences that result, directly or indirectly, still overwhelm the elasticity of the body and defy the laws of mutual aid. However, we have seen in the previous chapters how much the elastic human uses innate (genetic) operating loops acquired by themselves or their ancestors (epigenetics). The life of the elastic human is thus not only predetermined by their natural gifts, which would support the castes, patriarchy and representations of the difference. The human is not a simple machine that an education can direct, which would support meritocracy and the revolutionary logic of a clean slate to be rebuilt. A human is neither fundamentally selfish nor lazy by essence as some describe, which would support punitive education and moralizing laws. It is on this system of the elastic human that we can also base the moral sense of humanity and extend it to all living beings. For humanity also shares with other animals a physical, psychological and cognitive equipment, which does not allow it to boast of an existential biological superiority. Behavioral sciences even show that the notions of historicity and perfectibility attached to humanity since the writings of Jean-Jacques Rousseau can be extended to the whole animal kingdom, which singularly weakens the notion of an immanence attributed to humanity.

The process of artistic creation is ultimately a particular manifestation of the elastic forces of the human mind. Sensory-motor, mental and psychological resources are brought into play, jointly and alternately. It is a matter of stepping outside the box and inventing while respecting one's own imagination and the messages associated with one's own culture. Among these mental representations will be found symbols of history and mythology. Several emblematic figures of disability and illness have accompanied me during this work, which has constantly emphasized the difference (Korff-Sausse 2020). There are the dwarfs, the giants and all the characters with a bent skeleton. We have discussed growth problems in Laron syndrome and met people whose bone growth never stops. Williams-Beuren syndrome, with its partial deprivation of elastin, is described as elf syndrome because of the airy, smiling appearance of those affected. Blind Oedipus was constructed as a clubfooted child "exposed" to donation to find a foster family. Prometheus saw his liver devoured every day by an eagle to punish him for having given fire to men. Our Atlas vertebra carries our head like the titan Atlas carried the weight of the heavens since his defeat against Zeus. The anarchic calcification accommodates the myth of Medusa whose glance petrified. Isis reconstituted the exploded puzzle of Osiris' pieces except for the penis reconstituted with clay as biomaterial. Engineering for health is confronted with the wisdom of the ingenious Daedalus, who invented the eponymous labyrinth and wax wings to escape. It is his son Icarus who represents the unbridled increase of man's ambition melting his wax wings in his desire to get closer to the sun. Sisyphus became a slave of his rock as many human beings have become slaves of their technique. Finally, it was the broadcast of September 12, 2001, introducing a little girl with big black eyes, which provided me with the atmosphere for this book with the myth of the Apocalypse evoked during the attack on the World Trade Center. Modern times also have their mythologies to which the elasticity of the Living certainly contributes, by exploring our bodily limits and our representations of life.

7.4.5. *Morality and the elastic human*

One can be lost in front of the multitude of directions that seek to promote the relationship between humanity/nature/technique/culture. One can reasonably doubt that the cocktail of cross-linking – elasticity – resilience – compliance – conviviality – mutual aid – culture can save the world or at least protect it from our excesses. In reality, attempts and actions already exist under the pressure of a public opinion that is starting to become aware of the harmful effects of a non-convivial growth. It is time because these side effects are sudden and already very real. The extinction of multiple living species is effective, and the drastic modification of the climate is felt. The conviction of a "no-future" generation is taking shape. Several options are offered to the elastic human. Most of them consider a moral dimension.

An effective morality is often defined by a proper consideration of the opportunity cost. This concept corresponds to the choice one makes when moving from one situation to another. It is a matter of judging what one loses, what one gains and of gauging the effort required to accomplish the transition. This effort is often noticeable in the body when movement is required, even more so if the body is weakened, suffering or alienated. The transition dynamic may be stronger if the choice brings a real bonus. The situation is simple when the choice to accept a hip prosthesis is consolidated by the hope of being pain free and being able to walk again. The cost of the choice is then very low. On the other hand, the opportunity cost is very high when giving up the car also means losing a moment of listening in a confined and tight space, to make a longer, dangerous journey by bicycle dressed without any regard to the requirements of fashion. The concept also prevails when gluttony favors the intake of sweet foods to the detriment of an avoidance diet. We can generalize this concept to all the personal choices that we know are harmful, like those we have listed in this book, and that we do not abandon despite the scientific evidence, such as the use of junk food, cigarettes and alcohol, food from denatured agrochemicals, plastic, polluting energy or sedentary behavior. The opportunity cost can be taken into account for all actions that affect the elastic capital but that we will not change without rebalancing the benefit/risk balance. It takes a strong reward to change one's practices in a more respectful direction when a whole society encourages us to do the opposite. We have already dealt with this question of individual gain with the notions of conviviality and mutual aid. But this is clearly not enough.

Taking into account the opportunity cost is also the responsibility of national and local authorities. Their responsibility is great when they fail to consider the needs of the weakest with situations of disability or low autonomy, especially if they postpone decisions that are beneficial to society indefinitely. This laxity is cruel and ineffective because the solutions that are useful for the weakest are also useful for the strongest. The responsibility is also great when they do not respect the environmental values of the companies they manage. They could previously mask their inaction by presenting the burden of change as too sacrificial to be considered. But in the face of urgency, we must push a little. A constant, peripheral and directed pressure must be applied to avoid the constipation of society just as it is needed to advance the food in our intestines. Goodwill, mutual aid, technology or social and environmental empathy contribute to this pressure. But experience shows their low degree of effectiveness, or at least their slowness. A compliant justice at the service of conviviality could be effective in accelerating the movement. In any case, it is only this that has made it possible to change the course of stories that were bound to be lost in several public health scandals.

The effective recognition of the damage observed requires the introduction of a form of justice that challenges personal, corporate or societal responsibilities. These legal actions must generate and support the rare flow propagating from the weak to the strong. Justice is made for this, as long as it is balanced. It can enforce compliance, support resilience and include factual networks. It must be said, however, that it is only fairly recently that judgments about the moral obligations in public health of policymakers and corporations have become an option that can effectively combat other imperatives. Examples abound of conflicts between economic forces, the proponents of a reasoned ecology and individuals and patient associations. Examples abound of whistleblowers who pay a high price for their moral behavior.

During a CNRS workshop on law and disability that we were facilitating, Valérie Siranyan and I described as "Handicracy" a society that would be based on taking into account the inequity of situations of disability and loss of autonomy, regardless of their nature (Dussart et al. 2019). In relation to the theme of this book, we can consider all people with a restricted elastic capital, such as children, pregnant women and aging people who can also at any time be placed in a situation of inferiority. This concept of Handicracy brings together justice and the law with the convivial management of health introduced earlier. Battles have been won, but much remains to be done. Perhaps we should imagine a #caretooelastic movement that explores all situations where public health policies are not up to the task of protecting our precious elastic capital? The identification of the causes leading to pathological consequences on human elasticity, as outlined, can contribute to this inventory.

7.5. Chapter summary

Figure 7.2. *The elastic human's mind. The intention of the seventh painting is to present the harmonious immersion of the elastic human in nature and the environment. The dominant theme comes from the expression of some Nabis[1] like Paul Sérusier and his painting* Women at the Source. *Nature is present with two aspects, favorable or unfavorable to the elastic human. The deployment of balanced plant coverings or bucolic landscapes are favorable to the rituals and the appeased behaviors of the elastic human. However, the natural violence illustrated here by a representation of a Vesuvius releasing its network of fire and fluids illustrates the constraints of the mineral on the Living*

The real intention of this chapter was to test a hypothesis extending to the society the reading grid applied to the description of the elastic system in general, and of the body in particular. The vision was to add a layer to the already well-documented frameworks of philosophical, biological, sociological, psychological, legal, technical and cultural systems. At the end

---

1 Les Nabis were a young group of French artists in Paris from 1888 to 1900. They played a major role in the transition from impressionism and academic art to abstract art, symbolism and other forms of modernism.

of writing this chapter, I would like to suggest that the hypothesis deserves attention. More than all these other frameworks, the elastic system bears witness to a discreet, efficient but relatively poorly identified nature. As much as all these other systems, it is put to the test by the expansion of an energy-intensive, unequal and acculturated humanity. Like the other systems, it can benefit from laws to make itself recognized and to respect the balance of the quartet humanity/nature/technique/culture indispensable to a balanced society. Associated with the other systems, it composes the structured, fluid, pragmatic and sensitive framework of knowledge and learning. Like them, it obeys the laws of its components derived from biology, physics and chemistry (elasticity, cross-linking, resilience, compliance, conviviality; I could also have included the second law of thermodynamics on entropy). Converging with them, the system of the elastic human participates in the mutual aid and in the construction of a humanistic culture, which is and will be necessary to build a welcoming world for the future generations. As we envisaged with Valérie Siranyan in our seminar on Handicracy, we could imagine a society that is built on values that respect the elastic system: a compliant and convivial society respectful of the elasticity of the living, whether it is that of the human or that of the biodiversity.

Conclusion

The systemic presentation of the elasticity of living organisms was initially intended to present a capital that is relatively ignored because it is invisible for the most part, apart from the evidence of the skin. However, it is quite easy to identify the elementary notions concerning the brain, hormonal or immunological systems, skeleton, muscles, large systems (respiratory, cardiovascular, digestive) or reproductive organs, but it is not commonplace to identify elasticity as a complete system. This difficulty of representation is logical because the functions covered by the elasticity of living organisms are often hidden. While skin aging is easily associated with a loss of elasticity, the link is much more difficult to make with respiration, sensoriality, energy production and waste management. I hope that this book has provided insight into the few pieces of the puzzle of this complex capital with intimately related components, or at least to make people understand how the appearance of skin wrinkles can reflect a risk in any other part of the body.

The journey around the mechanics of the body, with elasticity as an focal point, has taken us far away from the classical comments on some kind of analogues of hoses, connecting rods and biological engines. On the contrary, elastic capital is a strange, scattered and difficult-to-identify capital that melts inexorably and has little possibility of renewal. On the other hand, it is a biological system, part of which is visible or sensitive. We can measure the state of its aging by the depth of its wrinkles, its breathlessness, the stiffness of the spine or its presbyopia, whether visual, auditory or even olfactory. Moreover, it is one of the systemic markers whose aging cumulatively measures the decline of body functions. We can also wonder about the link between the decline of a human being's elasticity and other organic functions, including the mind, because the harmonization of faculties is favored by a mechanically optimized body.

A holistic presentation of the elasticity of living organisms as a system has been made possible by a favorable scientific context. Recent research has indeed identified most of the molecules that shape the fibers that ensure the mechanical functions of living organisms. The identification at the beginning of the 21st century of the genes involved and their mutation causing elastic dysfunctions provides the basic knowledge for this book. It is the studies on imperfect elastogenesis that have made it possible to characterize precisely the functions of these complex and subtle amalgams that confer their properties to tissues. The elasticity of living organisms is indeed first of all a specific elasticity of tissues. Various assemblies are required according to the tissues considered. But it is their whole which will compose the general body mechanics and condition the physical potentialities of humanity.

Differences in elastic endowment are naturally a source of inequity. This inequity is severe for children who are not endowed with this initial potential, as a result of genetic diseases or unfavorable growing conditions, but it is also real between most men and women. The causes of these differences are often genetic. The skin elasticity of African, Asian or Caucasian populations is not equivalent for genetic reasons. It is also not equivalent for epigenetic reasons, when the living conditions appear different. Education and sports practices in childhood and young adulthood are essential to build up the full potential of an elastic capital, part of which is innate. This can be considered good news because it is possible to act on the living conditions without acting on the genetic causes, at least for the moment. We would like to think that the elasticity of living organisms can benefit from healthy living conditions, as has been observed in the length of telomeres.

Technologies for health and agronomy had already become a fundamental axis for the development of care, health, food and life paths. Experimental medicine has continued its impressive momentum. It can detect, diagnose, correct, clean and to some extent replace. It addresses the many facets of the structures and functions that contribute to the balance of our body (so-called homeostasis) and its post-traumatic adaptations (so-called allostasis, when for example the body integrates a knee prosthesis). It mobilizes natural, innate or acquired defenses (such as immunity), repair mechanisms (stem cells, blood platelets, control of DNA, RNA, telomeres and proteins) or chemical control (hormones and mediators). However, an overview has yet to be developed in the specific case of elasticity in the living world, as techniques for the identical replacement of elastic tissue or for its stimulation in reconstruction processes are not yet available. Studies are underway and offer hope. In the meantime, solutions exist but need to be improved.

When medicine based on experimentation proves to be powerless and no alternative treatment really exists, as is the case in the treatment of fibrosis and loss of elasticity, it is clear that people in pain turn to complementary approaches. Prevention through diet and exercise as well as traditional and ethno-cultural medicines are being used, which I had to testify to in this book. Thus, training elasticity through uninterrupted mobility helps to slow down the elastic decline. Fortunately, these practices are beginning to converge by adopting a scientific validation and a common vocabulary, to the great benefit of users who will be able to embrace a systemic care and prevention offer. When a drug is not available, ethno-phytotherapy can really be considered if it is first supported by fundamental research. It is the encouraging results obtained in preclinical (i.e. animal) trials that have brought dill extract into this story. Science will perhaps identify an active molecule later on, which could duly be added to the pharmacopoeia. Science advances in this way most of the time, in small incremental steps. This book has evoked several of these routes to deal with this strange capital whose tendency is to decrease. The research is going well and is in the hands of talented and motivated researchers. They will write the next biological and medical chapters of history. But it is also a matter of individual and collective responsibility to take care of this capital. A thousand discoveries are still to be written and described, supported by the strong desire to remedy the loss or the decrease of the elastic capital.

The times of pandemics are striking examples of the dynamics of scientific research mobilized by a major public health issue. The pandemic highlights the difficulty of understanding a pathogen's propagation mechanisms and, as a result, of producing an effective drug, single or multiple (as for the triple therapy to control AIDS) capable of effectively controlling its virulence. At the same time, studies that were not a priority, although never subordinate, can turn out to be dominant. The most fascinating example has been the acceleration of the use of RNA-based medicines, developed with exemplary speed for certain vaccines. Research on organ fibrosis is once again becoming a priority, whereas it had sometimes been somewhat neglected. Colleagues working on hearing, vision, olfaction, gustation, tactile and proprioception have seen a sudden and justified revival of their research. Emphasis is also placed on the control of blood pressure, glycemia, pathological obesity or autoimmune inflation, among others, which have suddenly become co-morbidity factors. It is interesting to note that these well-established biomarkers in public health all appeal in part to the elasticity of life. Let us bet that we will find these markers in long-term studies on current infections that are beginning to mobilize academic research, and rightly so.

Coronaviruses are certainly very old tenants of humanity. But their current virulence definitely marks the beginning of this century. At the same time, another imprint sees its spectrum growing. It is epigenetics that has regained the place it should never have lost. It revisits with benefit the discussions on adaptive evolution conducted by the great scientists of the 18th and 19th centuries. The weight of the environment (nature or society) is duly reinforced. This means that any action that alters this environment can affect our common history. It also means that not everything is defined by genetics a priori, which opens up very interesting medical and behavioral perspectives. We have thus been able to report, for example, that a genetic mutation affecting a protein involved in the elaboration of elastic fibers can have consequences on the expression of other genes, for good or bad, which is not insignificant in therapeutic terms. More generally, epigenetics introduces a broad spectrum of environmental responsibilities. It reminds us that humanity is only a tenant of an earth that it must respect; otherwise, it will face increasingly cruel difficulties as a result of ecological disturbances.

The writing of this book was largely motivated by the encounter with otherness. New friends inspired me a lot during the course of the study, whether they were patients, caregivers or researchers concerned with cutis laxa, pseudoxanthoma elasticum, Marfan or other Ehlers-Danlos syndromes. They taught me their relationship to difference and made me feel that their rare diseases were part of a general problem, that of the elasticity of life. This problem coincided with research on aging, since we all witness the programmed decrease of this elastic capital with age. It is as if each human had at birth an endowment which they would dispose of according to their possibilities, requirements, desires and the constraints of their physical, social and cultural environment. The question has more recently been made more complex by COVID-19, which has panicked the biomarkers of body mechanics, such as blood pressure, respiratory capacity, joint stiffness or even digestion. Some of the symptoms closely or remotely associated with the loss of elasticity are very visible, while others are invisible and make it difficult to notice the difference. Even more so if the causes are not determined or the consequences fluctuate. It is this precise level of questioning by those concerned that necessitated the expansion of the subject to the social and societal sciences, whereas it was previously limited to the fields of physics, biochemistry and biology.

The inexorable and irreversible decline of capital, whether economic, cultural or biological, suggests a political narrative underlying a societal program in which each human would have equitable resources at the beginning of life. In the case of the elasticity of life, equity is about compensating for initial or acquired differences, starting by identifying the causes of inequity in order to be able to develop solutions. It is therefore first of all a question of acquiring fundamental knowledge, the sum of which progresses at a pace that may sometimes seem very slow, especially when it comes to the complexity of life. However, the experience of pandemics and the real consequences of global warming on the elasticity of the living body show how much the notion of urgency can accelerate this pace. Let us hope that research will not slow down this pace to deal with the long-term effects of this virus, and other viruses at the same time. But it also remains to consider the sharing of this knowledge, which is so essential for the people concerned or simply worried.

Knowledge sharing has experienced a huge leap forward through social networks at the beginning of the 21st century. It is obvious that the maturity of these networks requires multiple adjustments. The experience of a highly publicized global pandemic may facilitate this transformation. The world has indeed experienced the anxiety that all researchers experience when developing knowledge (which viruses, which drugs?) and its applications (which remedies, which precautions?) about a deadly constraint (SARS-CoV-2, etc.). Everything and its opposite has been said and the bar room opinions have incubated like rancid exhalations. But at the same time, regulatory mechanisms are being built, and the identification of information sources is becoming paramount. After all, the construction of universal knowledge and its diffusion to all audiences is of the same magnitude as the creation of the *Dictionnaire raisonné des sciences, des arts et des métiers* by Denis Diderot and Jean le Rond d'Alembert, between 1750 and 1772. Its publication fueled a century of "Enlightenment" that still shines to this day.

We will hope that the consolidation of a shared knowledge on social networks will be in the same line. In any case, the sharing of knowledge during the pandemic has advanced the global understanding of the body as a complex entity where everything is linked to everything. A microbiological infection that targets "at the same time" such structured systems as the pulmonary, cardiovascular, renal, cerebral or sensory systems explodes the boundaries between specialties. People suffering from cutis laxa (e.g. but the list of diseases concerned is long) are familiar with this "simultaneous" pathology. The little girl on the reality show has grown into a fully grown adult. She looks a bit like her mother, who admittedly does not look her age. She has had, is having, and will have painful medical episodes that have taken her through many specialties. She fought to find a job and to take her place in civil society as a woman like any other. Her enthusiasm is extremely communicative, despite lungs severely impacted by emphysema and a necessarily limited energy. Her mother is on the front lines of rare diseases, including cutis laxa. She and her thousands of peers participate in the work of caregivers to relentlessly raise awareness of these particular diseases that escape the media's attention but whose number of affected cases calls out to the general public and imposes its political injunctions.

Political reflection is progressing to build a just and equitable society, respectful of itself, its members and its environment, but certainly in too small steps that are not taken at the same pace in all corners of the world. Palliative solutions have been and will be found, but political and legal decisions remain short of intentions when institutions do not sufficiently take into account otherness and our place in biodiversity. Several reading grids can be envisaged to respond to public and environmental health injunctions. In this book, we have introduced the draft of an original grid formed by the laws of mechanics and chemistry relating to living organisms. Indeed, the notions of elasticity, compliance, cross-linking (and network) and resilience are already used in everyday language, whether on an economic, social or societal level. This could be the starting point for a contribution to the development of solutions that respect the elasticity of humans in their environment. Among these solutions, some of them raise very worrying ethical questions since they introduce genetic modifications of living things. Their power is such that they require the systematic inclusion of legal safeguards to ensure that science is carried out in good conscience. The vigilance of whistleblowers, whether academic or individual, must be paramount in order to set the parameters for compliance by societal actors.

In conclusion, this book aimed to identify and expose a vision of the elasticity of living things, restricting itself to humanity. The presentation of the knowledge of this field proved to be very rich but complex, especially since it is under the pressure of a world pandemic and major climate change. To reinforce the memorization of this notion, I shared seven digital painting creations whose meaning went beyond the creation process, especially since they are inspired by famous painters. Many shores remain to be investigated that multiple explorers are currently criss-crossing. In short, this gesture of the elastic human perpetuates the practice of moral fables, like one of Aesop's Fables, *The Belly and the Members*, when the description of the human body illustrates morality and uses nature as a support. I hope that this book will have made better known this diffuse organ disseminated throughout the body, of which we must be aware and take care because most of our organic functions depend on it.

Appendix

A.1. Introduction to elasticity modeling by Professor Yves Rémond

That elasticity is a property of inert matter has been known since the dawn of time. It was only conceptualized in the middle of the 17th century, because of a great English scientist named Robert Hooke. Then, the engineers and scientists who followed continued its study, and their results are the basis of all designs of objects, machines, buildings or vehicles that surround us. Its extension to the observation of living things is however much more recent. It is striking to see that this Oxford scholar, to whom we owe the first acoustic telephone, had already synthesized the elastic law of bodies with a sentence that has become famous: *Ut tensio sic vis* (elongation is proportional to force). It is also interesting to note that Robert Hooke was also a pioneer in biology. We owe him the name "cell" for living plant cells, which he was already observing under the microscope!

Today, the elasticity of bodies is much better understood and modeled, and different forms can be distinguished. It is of course necessary to start from Hooke's famous law, which links the state of stress to the state of strain of a body in a point of matter. This law is expressed today by the following tensor relation (for isotropic media in any case):

$$\varepsilon_{ij} = \frac{1+\nu}{E} \sigma_{ij} - \frac{\nu}{E} \sigma_{kk} \delta_{ij}.$$

It can be noted that this isotropic elasticity requires the data of two independent elastic constants, Young's modulus E in Pascal that characterizes the medium's stiffness, and Poisson's ratio ν, which indicates the lateral effects during an elongation. Robert Hooke had stopped at a one-dimensional vision, which we find here when $i = j = 1$, so $\sigma_{11} = E\varepsilon_{11}$. The mathematical tools of the time did not enable doing better.

Biological media all have an elasticity, even if reduced, just like inert matter. When mechanical behaviors are linear, this elasticity is described by the same Hooke's law recalled above. However, these biological media are rarely isotropic. More constants are needed to characterize them: five for transverse isotropic media, a venous or arterial wall for example, nine for orthotropic media. Living nature has not gone so far as to propose weaker symmetries leading to monoclinic properties as found in crystallography requiring 13 elastic constants, or even totally anisotropic properties as with some rocks whose elasticity requires 21 elastic constants to be described.

Even more striking is the distinction between two kinds of elasticity, fundamentally different from the point of view of thermodynamics. For inert media, a simple experiment allows us to separate them. It involves stretching a piece of material (a tensile specimen), then blocking the strain thus acquired and finally heating the stretched material. We then observe the evolution of the shrinkage force during this heating, force that the tested material sample exerts on its attachments. If this force decreases as the temperature of the material increases, then it is enthalpic elasticity; if it increases, it is entropic elasticity. In the enthalpic case, the equilibrium state corresponds to the medium's minimum enthalpy. In the entropic case, the natural state of the material concerned corresponds to its maximum entropy; any elongation will reduce this entropy, but the force of shrinkage comes from micro-Brownian movements that increase with temperature. And in the case of living tissues you may ask? We have no precise information. It is probable, as in the case of certain polymers and elastomers described by these two types of elasticity, that we are in the presence of a mixture of these two types of elasticity, but the question remains open, and everything will probably depend on the scale on which we will place ourselves.

Finally, let us also remember that the material offers us interesting variants of elasticity.

First, there is viscoelasticity that introduces delays in the elastic reaction of the solicited material. This is typically the case for living matter, where viscosities are at work at all scales. We can measure viscosity maps as well as stiffness maps of living tissues and derive useful clinical information from them. We can be interested in this for muscular tissues with regard to Duchene muscular dystrophy for example, or for the rigidity of tumors.

We could talk about superelasticity. This is typical of certain artificial materials used in the human body for their ability to have a shape memory. It is used for example to manufacture stents, orthodontic wire, etc. This particular elasticity comes from a phase change of the material during solicitation. This phenomenon is then used to store particular deformations and restore them, thus activating this memory effect. The most classical inert materials in this field are nickel titanium alloys.

Finally, let us mention hyperelasticity, a concept that has allowed us to understand and model the nonlinear elastic behavior of many media, especially biological ones. In this case, we use the data of a thermodynamic potential like the free energy from which the law of behavior is derived, to characterize them.

A.2. Introduction to the actors of the elasticity of life

A.2.1. *The extracellular matrix*

The extracellular matrix supports tissues and organs, and it conditions their structure and mechanical properties while just informing the cells. Its structuring is evolutionary between development and aging. Its physical characteristics are linked to the organization of fibers, to the contractility of cells and to the interaction between these cells and the network constituted by the extracellular matrix. Its composition is specific to each tissue, with a variable composition and arrangement of the families of components, which are as follows:

– elastic microfibrils and elastic fibers that give elasticity;

– collagen fibers that provide strength, support and viscoelasticity;

– proteoglycans (with hyaluronic acid) that provide viscoelasticity and compressibility;

– glycoproteins that support and maintain structures, such as the basal structure at the interface of the dermis and epidermis;

– intracellular and muscular filaments that enable tensile integrity and support contraction and extension;

– (lipid-based) mucus that participates in the surface tension.

Thus, composite and reticulated filaments, structures and assemblies ensure the mechanics of the body. The solidity of the skeleton is a function of the collagenous network and its adapted calcification, and the flexibility and elasticity are a function of the network of fibers (elastic, collagenous, muscular filaments) and their composition, as well as their hydration which is essential to ensure the elasticity of the fibers.

A.2.2. *Elastic fibers*

Elastic fibers are essential components of the extracellular matrix. They allow the properties of extension and elastic recoil characteristic of tissues whose physiological role requires these repeated distensions and relaxations throughout life. But they are never or rarely alone, and the whole matrix must be taken into consideration.

The organization of elastic fibers is specific to each tissue but is always composed on the basis of microfibrils on which elastin can be affixed (Figure 1.3). Microfibrils are found in primary organisms such as jellyfish, invertebrates and vertebrates, whereas elastin appeared more recently in evolution to reinforce the tissues of the circulatory system of higher vertebrates. Elastin-free microfibrils are also found in the ciliary body of the eye, where they play an important role in the dynamics of the eye and lens in dynamic suspension (Ali et al. 2009).

The core of elastic fibers is therefore composed of elastin, a polymer made up of soluble monomers of tropoelastin, a major player in fiber elasticity. In addition to elastin, many other components make up elastic fibers, including fibrillins, fibulins and glycoproteins associated with microfibrils.

The formation of elastic fibers begins early in development and continues mainly until the end of the growth phase. Their assembly is very complex, and they are made up of close to 30 proteins (Baldwin et al. 2013). Schematically, their synthesis first requires the production and extracellular organization of microfibrils, of which the major proteins are fibrillins. The cells will then deposit elastin on this framework, in a very dynamic way and controlled by numerous proteins such as fibulins and lysyl oxidases (LOXs). Observed by transmission electron microscopy, the elastic fibers then appear as intertwined fibrils, on which an amorphous core of cross-linked elastin appears. The fibers are about 100 nm in diameter, compared to 3 nm in diameter for the filaments. In the stretched state, the distance between the two filaments is 5 nm.

Tropoelastin monomers combine to form aggregates with fibulins 4 and 5 and also carry immature lysyl oxidases. Hydroxylation of the prolyl groups of elastin leads to secretion by Golgi vesicles into the extracellular matrix where LOXs are activated. The aggregates align along the microfibrils. The active LOX enzymes initiate bridging between elastin monomers to create an independent elastic fiber through the microfibrillar structure that provides the assembly's architecture.

A.2.3. *Microfibrils associated with elastin*

The microfibrils form the "lattice" on which the elastin molecules will align. They have a diameter of 10–12 nm and appear in the form of strings of beads spaced 56 nm apart. Their main components are fibrillins 1 and 2. In the presence of calcium, these large glycoproteins (proteins associated with a sugar molecule) can change their structure and elongate. Fibrillin-2 is expressed in the early stages of development, whereas fibrillin-1, which is the majority, is expressed throughout life. The fibrillin molecules are assembled longitudinally in an antiparallel way. They are assembled in stable networks because of an enzyme, the tissue transglutaminase, which generates bridges between two types of amino acids of these proteins (glutamines and lysines). These linkages play an important role in the stability and elasticity of microfibrils, which have a certain proportion of extension (Kozel and Mecham 2019).

Microfibrils also contain many other proteins. These include microfibril associated glycoproteins (MAGP) 1 and 2. MAGP-1 is highly expressed during development and localizes to the beads of microfibrils. MAGP-2 is restricted essentially to development and to few tissues.

Numerous molecules participate in the interface between elastin, microfibrils, the extracellular matrix and cells. They regulate the deposit of elastin on the microfibrils and connect the elastic fibers to the cell surface. Fibulins are present at the interface with elastic fibers and are strongly involved in their formation and function. Thus, the absence of fibulin-1 weakens the vascular wall in animals that then die within 2 days after birth from vessel rupture and major hemorrhage. However, fibulin-2 deficiency introduces a hypoxic situation (with low oxygenation) and incomplete formation of elastic fibers (Boizot et al. 2022).

A.2.4. *Elastin and lysyl oxidases*

Elastin represents the majority of the proteins in elastic fibers. It was first isolated because of its great resistance to chemical products. It is also resistant to numerous enzymes that degrade proteins (proteases), which gives it, in its mature form, a half-life estimated at 70 years. These unique properties of elastin require a very controlled synthesis. Elastin is first synthesized in the form of a soluble molecule called tropoelastin. This precursor, itself elastic, is very sensitive to degradation (Wahart et al. 2019). It has to be protected by a companion molecule, then secreted by the cell in the form of a kind of microbeads and deposited on the microfibrils where, after transformation, it will become very stable and endowed with its characteristic mechanical and chemical properties.

The sequence of elastin instructs its main function, which is its capacity to expand and contract when hydrated. Elastin functions as a material in which the elastic recoil is directed by a return to a more favorable energetic (entropic) situation during relaxation after extension. Its sequence presents an alternation of hydrophobic domains (rich in the amino acids glycine, valine and proline) and more hydrophilic cross-linking domains (rich in proline and lysine). The hydrophobic domains are involved in the formation of extensible loops, which are the basis of the mechanical properties of elastin. In its soluble form, this protein behaves like a perfect spring, capable of extending up to eight times its original size. The cross-linking domains are the sites where covalent bridges are formed, within the molecule or between molecules.

The mechanical and chemical properties of elastic fibers also depend on stable bonds between elastin molecules and fibers, which is called cross-linking or cross-links. In other words, fibers would not be as elastic or as strong without the formation of this intra- and inter-fiber cross-linking. The first step in the formation of these cross-links involves lysines modified by a family of enzymes called lysyl oxidases. As their name suggests, these proteins oxidize lysines on the soluble precursor of elastin (tropoelastin). This initiates a complex chain reaction that results in the formation of a dense network of elastic material that is extremely stable and very elastic, associated with microfibrils. This cross-linking step is also found in the formation of collagen fibers, depending on the initial activation by lysyl oxidases (Vallet and Ricard-Blum 2019).

The family of lysyl oxidases consists of five forms. The first to be discovered was named LOX, while the following ones are called lysyl oxidase-like (LOXL1 to LOXL4). Their specific role is still under investigation, but they are essential for the genesis of collagen and elastic fibers. Animals with an anomaly of the emblematic LOX are not viable, which is not the case for those with a LOXL1 deficiency, which also presents multiple malformations.

The activity of these enzymes is dependent on copper, so it is not surprising that deficiencies in the metabolism of this compound result in a deficit in the formation of collagen and elastic fibers, among other often extremely serious manifestations found in the Menkes syndrome related to this copper deficiency. Associated proteins are also necessary, such as the presence of fibulin-5 which participates in the assembly between the soluble precursor of elastin, fibrillins and the lysyl oxidase LOXL1. Its absence in mice leads to disorganization and fragmentation of elastic fibers in all organs (pulmonary emphysema, loose skin, tortuous aorta, etc.).

It is the action of the lysyl oxidases that is inhibited by a toxin of the sweet pea (the *almorta*) with the deleterious consequences that we group under lathyrism (from the genus name of the plant *lathyrus*) and are represented in Goya's painting.

The activity of lysyl oxidases is frequently evoked for its role in cancers. The LOX protein (and the corresponding gene) was first identified from a biochemical and immunochemical approach, as the molecule responsible for the initiation of collagen cross-link formation. It turned out that the identified gene corresponded to the one coding for an inhibitor of ras oncogene-stimulated tumor transformation. In this context, the LOX gene and the so-called ras recision gene (rrg) were found to be identical, and LOX could carry antitumor activity. LOX protein expression was then associated with the formation of peritumoral tissues in non-invasive breast cancers, in contrast to invaded tissues where it is not present, suggesting a role in limiting tumor expansion by the reticulated stroma surrounding the tumors. And finally, in complete contrast, the presence of LOX was directly associated with tumor expansion and hypoxia accompanying the tumors, demonstrating a protumoral activity of the molecule (Reynaud et al. 2017). The molecule is currently under investigation in the search for anticancer pharmacological inhibitors. Research continues, but it is certain that LOX is considered an important target in cancer research.

A.2.5. *Collagen fibers and rigidity*

Collagen is a fibrous protein and plays an essential structural role. It is the most abundant protein in the body. Contrary to elastin, collagen is difficult to stretch and resists traction well. There are about 30 different types of collagens, of which type I collagen is the most abundant. These type I molecules are made up of three polypeptide chains wound together to form a triple helix. This helix is stabilized by a network of intramolecular hydrogen bonds. These monomers are wound together into protofibrils and then into fibrils that appear as a tight rope with a characteristic regular striation of 67 nm in mammals.

Several studies have been carried out to characterize the mechanical properties of these fibrils. Collagen fibrils behave like a viscoelastic material in which water molecules are exchanged between the hydrogen bonds, thus leading to energy dissipation. The elastic portion of the viscoelastic behavior represents the reversible stretching of the triple helices, and the viscous part reflects the viscous friction between fibrils. Cross-linking limits this slip and increases the elasticity.

A.2.6. *The actors of viscosity*

The fundamental substance of tissues embeds the fibrillar network as well as the cells. This high-water-content gel contains glycosaminoglycans

(GAGs), proteoglycans and structural glycoproteins. These molecules soak water into the extracellular matrix, which regulates hydration and resistance to compressive forces. They are involved in the viscosity of all support tissues, including the skin by promoting the sliding of the fibers.

GAGs are composed of specific disaccharide units, glycosamines. They have a strong hygroscopic power, capable of binding up to 1,000 times their weight in water, which gives them a crucial role in skin hydration. They are divided into two groups: sulfated and non-sulfated. Among the non-sulfated GAGs, the most abundant is hyaluronic acid. This influences the viscosity and permeability of the extracellular matrix according to its degree of hydration. Among the sulfated GAGs, the main ones are chondroitins, keratans and dermatan sulfates.

Proteoglycans are polypeptide chains to which glycosaminoglycan molecules are attached. They are sensors of cations (Na^+, K^+ and Ca^{2+}) and water, and they allow the mobility of molecules within the extracellular matrix and are thus involved in the activity and stability of signaling molecules. The skin contains many proteoglycans including versicanes, derived from chondroitin sulfates, and decorins, derived from dermatan sulfates and biglycans, close to decorins but with two GAGs. These molecules are found in specific areas of the tissues and are able to interact with collagen and elastic fibers.

A.3. Elasticity and the structures of the body

A.3.1. *Elasticity and Young's modulus*

Like all bodies, it is possible to evaluate an elasticity of living materials. We evaluate empirically this elasticity by palpation or chewing, the skin being more elastic than bones and less than tendons and ligaments. The elasticity of these biomaterials is, however, measurable, even though they are generally complex composite materials. The scale shown in Figure 1.4 is indicative of the relative values of different tissues as assessed by their elastic modulus. This value, called Young's modulus (and expressed in Pascal), relates tensile or compressive stress with the onset of strain of a material (as measured by its elongation, bending or strain). The stiffer the material, the greater the stress (force) needed to deform it. Steel and diamond have a very high modulus, while plastics and organic materials do not. In the body, bones have the highest overall Young's modulus and will appear rigid, and they will deform less than ligaments and tendons, which have the best overall elasticity in the body, in any case much more than soft organs like the liver. Stiffness adds the notion of section, a massive bone being stiffer than a thin bone. Lastly, the hardness of skin and bone is obviously different, with the surface of the bone offering much more resistance than that of the skin (it will be necessary to press much harder to deform the bone given the resistance of its surface).

A.3.2. *Elasticity, skin and the musculoskeletal system*

The skin can be taken as an example to represent the notions of elasticity and its evolution, since we are all very conscious of its modifications. The skin is a multi-layered composite material composed of an upper cellular layer, the epidermis, 60–1000 μm thick, connected to a lower layer, the dermis, made of cells and an extracellular matrix, 1–4 mm thick. External forces are transmitted through the epidermis to the dermis and underlying tissues, whereas internal forces are transmitted from the dermis to the epidermis (Figure 1.2). External forces include shear forces from friction as well as extension and compression forces. Internal forces in the skin are passive tensions within the collagen fibers of the dermis directed approximately in the direction of Langer's lines. These are the tension lines of the skin located in the reticular zone of the dermis which are mainly due to the organization of the dermal fiber network. The drawing of the lines is influenced by various factors: morphology, body type, musculature, age and position of the subject. The active cellular tensions also follow the axis of Langer's lines. These forces are produced by the fibroblasts which contract the collagen fibers. In the absence of external forces, the internal tensions acting on the collagen fibrils of the dermis cause tensions at the junctions between the main cells of the epidermis (the keratinocytes). External forces applied to the skin surface increase these tensions and modify the state of tension of the dermis. External forces applied to the epidermis include normal forces resulting from compression or extension as well as shear forces resulting from friction.

The internal tensions of the dermis, in the axis of Langer's lines, are transmitted to the epidermis through an interface blade called the "junction" between the dermis and the epidermis. The mechanical continuity at this dermal-epidermal junction is the key to the transfer of internal and external mechanical forces, with essential constituents such as type IV collagen and laminins. The epidermis and dermis are connected at this junction through an attachment structure called hemidesmosomes. Anchoring filaments, composed of type VII collagen, are usually in high density in the hemidesmosomes. In genetic pathologies affecting type VII collagen, such as epidermolysis bullosa, the epidermis separates spontaneously from the dermis under the application of a stress. This observation underlines the importance of mechanical continuity between dermis and epidermis in order to maintain a normal transfer of mechanical forces (Thieulin et al. 2020).

The dermis is organized into two regions based on differences in density and arrangement of connective tissue. The upper layer, the papillary dermis, is about twice the thickness of the epidermis and is composed of fine, loose collagen fibers and elastic fibers. The deeper, reticular dermis contains large, tangled elastic fibers and collagen fibers.

The internal stresses of the dermis are related to the internal tensions of the collagen fiber network on the one hand, and to the tensions generated by the traction of the fibroblasts between them and on the collagen fibers and elastic fibers on the other.

To summarize, in the skin, the elasticity due to elastic fibers is essentially located in the upper dermal layers, especially at the interface between the dermis and the epidermis where they can take on a so-called candelabra appearance. The epidermis is in tension due to interactions between the keratinocytes. The dermal-epidermal junction is rich in specific proteins (laminins) and types IV and VII collagen fibers. The dermis is a connective tissue rich in collagen fibers and elastic fibers that ensure the maintenance and quality of the dermis (strength, viscosity, tensegrity). This tissue is crossed by multiple vessels (vascular and lymphatic) and nerve fibers.

Stiffness in bone and cartilage is essentially defined by the structure of the calcified collagen fibers of the bone and cartilage walls. Bone walls are formed of more or less cross-linked and calcified collagen fibers. Other structures are present, vascular and nervous, as well as a thin membrane, the periosteum surrounding the bone, whose composition is richer in elastic fibers. The knee joint is an example of the different mechanical systems involved. The bones are made up mainly of calcified collagen fibers; they are very rigid and hard. The cartilages and menisci are more elastic and must also be resistant to stress and pressure. Mobility and maintenance on the axes of movement are provided by ligaments between bones and tendons between bones and muscles. Muscular contraction ensures the dynamics. The continuity of the whole is ensured by the fascias (rich in collagen fibers with some elastic fibers). Resistance to friction in movement (tribology) and hydration (hydraulic tension) is facilitated by the synovial fluid rich in proteoglycans and GAGs such as hyaluronic acid for the best known. The nutritional and nervous connectivity is ensured by blood and lymphatic vessels, as well as by nerve fibers (to ensure proprioception in particular).

A.4. Elasticity and the circulation of fluids and gases

The elastic characteristics of the tissues that support the circulation of fluids and gases potentiate the function of the tissues, which are either at the interfaces between the exterior and the interior (skin mucous membranes and pulmonary alveoli) or are the actors of transport (muscles, organs and interstitial tissues) and the circulation of liquids and gases (lungs, diaphragms, heart and its four valves, vessels, and pleura to maintain a negative pressure).

A.4.1. *Blood circulation*

The composition and organization of elastic tissues are different throughout the blood circulation chain (Figure 2.3). The cardiovascular elastic system includes first the heart, with its elastic walls and valves. The large arteries, which are subjected to high pressures of the heartbeat, are iconic models to study elasticity. They have thick and very elastic walls with a reduced lumen. This center is surrounded by an internal wall (the intima) composed of endothelial cells leaning against a so-called basal lamella (membrane). The elasticity of the arteries is ensured mainly by a sheath of contractile cells within a dense and concentrically organized elastic tissue. The vascular smooth muscle cells of the media are embedded in a maze of collagen and elastic fibers. The outer wall (the adventitia), rich in collagen fibers, ensures the strength of the whole. The arterioles have thin walls with a reduced lumen, but they are contractile to regulate the flow by vasomotricity; the capillaries have thin walls with a reduced lumen and are not very contractile; the veins have thin, deformable and not very elastic walls, with a large lumen, which explains varicose vein problems.

Peripheral nerves are supported by elastic sheathing, blood and lymphatic vessels. The alteration of the elastic sheathing could be involved in chronic pain (fibromyalgia), especially in the spinal cord and peripheral extensions. The elastic system will also support the structures, extensions and ramifications of the peripheral nervous system as well as, to a lesser extent, the central nervous system via the sheathing of the cerebrospinal fluid. It will participate in the protection against compression, in the nutritional supply of the nerve cells by a vascular and lymphatic accompaniment. The cerebrospinal fluid (CSF) of the central nervous system is contained by the meninges, made up of a calcified dura mater and a more or less elastic arachnoid membrane. The CSF continues in the meninges from the spinal cord to the second sacral vertebra and participates in nutrient exchange.

A.4.2. Breathing

The lungs are a good example to demonstrate the essential role of elasticity in the body's energetics. Elasticity is involved in the structuring of the trachea, bronchi and bronchioles, and especially in the walls of the alveolar sacs (Figure 2.2). During a lung infection or air toxicity (pollution), the walls of the pulmonary alveoli, which are encircled by elastic fibers, can be very stressed (Figure 5.2).

According to the law of gas diffusion, the partial pressures of gases tend to equilibrium. Oxygen will spontaneously enter the body from the air, where it is more concentrated, to the body where it is less concentrated. The partial pressure of oxygen in the air is about 160 mm Hg (21% of 1 atmosphere at 760 mm Hg), whereas it is estimated to be less than 100 mm Hg in the body at the sites of consumption. The opposite is true for CO_2 produced by the combustion of oxygen and sugars; CO_2 will therefore tend to escape from the body into the air where it is less concentrated (about 22 mm Hg if present at 0.03% on average in the air, compared to an estimated 27–45 mm Hg in the tissues). The gases pass through the trachea, bronchi, bronchioles and pulmonary alveoli.

The constancy of pressure and volume balance is often invoked during inspiration and expiration. The extension of the diaphragm increases the volume of the lungs. This increases the inflow of O_2 (to restore the balance of partial gas pressures). Exhalation (passive) leads to a decrease in lung volume, which flushes out the CO_2 for a pressure rebalancing with the CO_2 of the air during its passage in the veins under the lungs. Convection of the blood under the effect of the heartbeat decreases the partial pressure of O_2 of the subpulmonary vessels, which promotes the transfer of O_2 from the lungs to the blood vessels. Conversely, the pressure of CO_2 produced by tissue activity will accelerate the transfer to the lungs, where it will diffuse into the air. The difference of affinity for O_2 and CO_2 of hemoglobin, the unique carrier of these two molecules, intervenes at the level of the transfers of these two gases, at the level of the cells and the tissues, when one replaces the other (O_2 will diffuse toward the cells and will be driven out of its carrier, the hemoglobin, which has more affinity for CO_2 produced).

Blood circulation enables the delivery of gases diluted in the blood to the sites of energy production. This circulation is controlled by the synchronized action of the two parts of the heart which manages the systemic circulation sending oxygenated blood throughout the body, while the pulmonary circulation pumps deoxygenated blood to the lungs. Diastole corresponds to the relaxation of the heart which fills up (with a pressure of about 80 mm Hg), while systole corresponds to its contraction. The pressure during diastole is about 80 mm Hg and 120 mm Hg during systole, which gives the famous figures of an optimal pressure of 12/8 (which is an average, when everything goes well).

The circulation of the body's energy is essential to transport the different elements necessary for the creation of adenosine triphosphate (ATP). As a reminder, energy comes from the breakdown of a phosphate from this adenosine tri-phosphate, formed by respiration in the presence of a carbon source. ATP is produced and distributed continuously, which requires a very efficient regulation of the circulation of the indispensable molecules and their product, as schematized in Figure 2.4. The ATP is formed in the cells, essentially in the micro-compartments of the mitochondria, either with glucose and oxygen (aerobic respiration), or without oxygen (anaerobic) with stored sugars or lipids and proteins and an 18-fold lower efficiency, or finally without glucose with fatty acids or proteins (ketones), at the level of the mitochondria.

A.4.3. *Water balance and transport of products generated by the body*

The circulation of energy sources and their waste products is in liquid and gaseous forms. The body is made up of about 60% liquids, all of which circulate. The inflow and outflow of water must be in balance (about 2.5 L/day), and the total liquid is renewed in about 50 days. Several liquid networks intersect, whether they are intracellular (about 24 L for a 70 kg person), interstitial (about 9 L), cardiovascular (about 5 L), lymphatic (about 2 L) or associated with sensory, nervous and digestive functions. Water balances therefore require perfectly orchestrated biomechanics. The water systems (blood, lymph, urine, transpiration, interstitial water and cerebrospinal fluid) are closely connected, and their composition is highly regulated because they define the hydrostatic, osmotic and oncotic balances. In short, this concerns nutrition, inter-organ connections, acid-base regulation, prevention of edema and effusions, waste elimination, regulation of potential differences and inter-body flows.

The regulation of pressures is orchestrated (a) by gravity and the pressure of the atmosphere (1 atmosphere = 760 mm Hg), (b) by the mechanical pressure due to the motor behavior of the muscles (heart, muscles around the organs and vessels), (c) by the elasticity of the ducts and containers (intestines, lungs, bladder, blood vessels and lymphatic vessels), (d) by the volume and flow of water, the composition of the fluids and their protein (osmotic pressure) and salt content (hydrostatic and electrolytic pressure – sodium, calcium, phosphate, iron, magnesium and acidic ions -H+) and (e) by the elasticity of the components (itself dependent on a good content of bonded water!). The regulation of the pressure is controlled (a) by the kidneys (in particular by the renin-angiotensin hypo- and hyper-tensor system), (b) by the diuretic hormone (kidneys and pituitary gland), (c) by water retention (kidneys, adrenals and the liver) and (d) by the sympathetic and parasympathetic systems (heart and vessel pressure).

Since the kidneys are mentioned, it is important to know that renal mechanics is very dependent on the organization of collagenous and elastic tissues. Filtration takes place in glomeruli formed by tubular and spherical structures with 1 million nephrons, i.e., 80 km of tubes with a strong presence of collagen and elastic fibers. For the record, the role of the kidneys is to ensure the filtration of blood and the elimination of waste, the control and the elaboration of urine (1–2 L/day), the reinjection of the filtered blood in the vascular circuit (200 L of blood/day) and thus the constant maintenance of blood volume (volume), the control of blood pressure, oncotic pressure (proteins), plasma tonus (Na+ natremia and glycemia) and the maintenance of the acid-base balance (pH), together with respiration, through the production of bicarbonate.

Waste is eliminated by the intestinal, urinary, lymphatic and sweat systems, all of which require physical mechanisms to function effectively. Several waves push the food bolus through the intestines. The wall of these viscera is made of an epithelium rich in villi. This mucosa is surrounded by muscular structures, hence the name muscularis. The elastic impregnation of peripheral tissues of the digestive system (esophagus and intestines) on the muscularis and their elastic and collagen fibers support the autonomous muscular movements (peristaltic) and structure the intermediate valves (cardia – gastroesophageal junction). Several waves run through these viscera. Contraction waves ensure segmentation and mixing; progression waves, decreasing from top to bottom, push food and residues; slow waves with synchronous ring contractions (small intestine) or asynchronous non-ring contractions (colon) support transport.

The general organization of the urinary system includes the kidneys, the ureter, the bladder and the urethra, all of which have an elastic behavior; the bladder, in particular, must be able to fill and empty itself thanks to a cellular structure enveloped by muscle cells that are associated with a stretchable matrix of collagen and elastic fibers.

The lymph is collected in the initial lymphatic vessels. It circulates in the pre-collectors and the contractile collectors equipped with elastic anti-reflux valves (with about 10 contractions/min) then in the lymphatic nodes (superficial or deep associated with the organs) to join one of the two lymphatic trunks (right and thoracic) and ends up in the venous circulation (jugular veins). Lymphatic channels have recently been discovered in the brain, where their role appears to be important for the elimination of toxins, especially during sleep.

A.5. Motor and autonomous sensory receptors

Mechanics and elasticity have a role to play in the eyes (ciliated bodies, lens, Bruch's membrane and optic nerve sheath), ears (tympanic membrane, ossicles, basilar membrane, oval window, cochlea, eustachian tube, etc.) or olfaction (nasal turbinate, walls of the nose, etc.). An overview is presented in Figure 3.2, while the concept of elastic sheathing of nerve tissue is introduced in Figure 2.3. Several types of receptors are permanently at work to instruct the body about its mechanical solicitations. The proprioceptive mechanoreceptors are related to conscious sensations, unlike the autonomic receptors. Almost all of them are related to the mechanics of the body's solids and liquids. They are essential for our lives. We distinguish:

– The conscious proprioceptive (position, extension, kinesthesia and forces) and autonomous mechanoreceptors found, for example, in the tendons, muscles, ligaments, pharynx, mandibles and the vestibular system.

– The cutaneous receptors which are involved in touch, recognition of pressure, texture and shape. They are found in particular in the skin, mucous membranes and fascias.

– The baroreceptors (voloreceptors and chemoreceptors) monitor blood pressure and the pressure of O_2, CO_2 and acidity. They are found, for example, in the sinuses, the aortic arch of the carotid artery, the cardiac atria and associated with various nerves (Hering and Cyon nerves).

– The peripheral osmoreceptors (osmotic tonicity) measure tension and tonicity. They are found in the esophagus, pharynx, intestines, vessels (hepatic portal vein, splenic and mesenteric vein), ears and kidneys (juxtaglomerular apparatus).

– Thermoreceptors will indicate the information necessary to initiate vasoconstriction or vasodilatation. They are found in the skin, vessels, esophagus, stomach and abdominal veins.

– Nociceptors (pain) are present in the internal organs and the external tissues.

A.6. Pathologies due to a deficiency of the elastic system

The loss of elasticity concerns everyone, since it is a factor that decreases with aging. However, it can be caused by multiple causes and accelerate in many situations when the repair or regeneration capacities of the tissues are altered.

The causes of loss of elasticity concern:

– structural elements: tendonitis, fasciitis, bone fragility, cartilage wear (menisci), hernias, etc.;

– reinforcement of structures: pathological calcification and cross-linking;

– fluidic (liquid, gas) and energetic regulation: cardiovascular, pulmonary and lymphatic insufficiency, aneurysm, edema, emphysema, fibrosis and stenosis;

– digestive system: digestive insufficiency, diverticula, intestinal leakage, reflux, glomerulonephritis and hepatic fibrosis;

– sensory-motor and emotional control: blindness, glaucoma, deafness and anosmia;

– the perineal space and reproduction: prolapse, infertility, fibromyalgia and difficulty in urinating.

The causes of loss of elasticity and associated pathologies are as follows:

– malformations: genetic, epigenetic or traumatic causes;

– degradations: inflammation, oxidation, rupture, tearing, chronic alterations (arthrosis, arthritis) and insufficiencies (respiratory, vascular);

– compressions and extensions: Morton's syndrome, carpal tunnel syndrome, Osgood-Schlatter's syndrome, herniations (discs, inguinal), tears and ruptures of ligaments and tendons and weakening of the menisci;

– rigidifications: calcifying tendonitis, chondrocalcinosis, calcaneal spur, fibrosis, stenosis, ankylosing arthritis and vascular calcification.

The consequences of loss of elasticity are amplified when the capacity for repair and regeneration is weak or impaired:

– slow regeneration of the dermis and fibroelastic cartilage (elastic potential almost disappearing in adulthood);

– the permanent but very slow regeneration of bones and ligaments;

– the limited regeneration of elastic fibers in adults.

The importance of the spectrum of diseases associated with a deficit in the formation and function of elastic fibers is demonstrated by the inherited diseases caused by mutations in the gene coding for the components of these fibers. Diseases of the elastic tissue combine disorders of mature elastic fibers and disorders of the microfibrils forming the base of mature fibers. Defects in the formation of collagen fibers are also associated with elasticity problems. These diseases are rare and most often of genetic origin, although forms acquired in adulthood, after a severe inflammatory episode for example, are reported more and more frequently.

Mutations affecting the major components of elastic fibers often lead to severe pathological disorders. This is true for mutations affecting all or part of elastin, fibrillins and fibulins 4 and 5. It is also true for mutations affecting lysyl oxidases and the metabolism of copper on which they depend. However, new mutations have recently been discovered in genes that did not appear to be directly related to elastic fiber formation, such as a multidrug resistance (MDR) transporter or a gene involved in glycosylation. Figure 4.2 provides a (transient) review of the state of knowledge about genetically based structural defects in the elastic system (Beyens et al. 2021). Let us look at some of these syndromes in more detail.

A.6.1. *Cutis laxa, supravalvular aortic stenosis and Williams-Beuren syndrome*

Cutis laxa ("loose skin") refers to a disease with loss of elastic fibers in the skin and connective tissue. It is certainly the most emblematic manifestation of disorders affecting the function of elastic tissue. It is a very heterogeneous disease in its clinical manifestations, which can be of genetic or acquired origin.

The autosomal dominant form of cutis laxa (ADCL) is very rare and is considered a moderate form. It has been associated with mutations in the elastin gene or, more rarely, in a gene coding for a mitochondrial enzyme (ALDH18A1). It is characterized by predominant skin involvement but can also present more serious risks such as aortic aneurysms and pulmonary emphysema. All patients have a defect in the formation of elastic fibers.

Autosomal recessive forms are the most common, with a spectrum ranging from a simple wrinkled skin syndrome to severe developmental and growth abnormalities. Autosomal recessive forms of cutis laxa (ARCL) have, in addition to loose skin, pulmonary (emphysema), cardiovascular, and even skeletal and neurological implications. ARCL type I has the worst prognosis with a fatal outcome in childhood. It has been associated with mutations in fibulin-5, fibulin-4 or latent transforming growth factor beta binding protein 4 (LTBP4), which is associated with fibrillins (Beyens et al. 2021).

ARCL type II, more common than type I, is associated with joint hyperlaxity and developmental delay. This type is related to mutations in proteins involved in cell trafficking (the V-type H+ ATPase subunits ATP6V0A2, ATP6V1E1 and ATPV6V1A) and a mitochondrial enzyme (pyrroline-5-carboxylate reductase 1, PYCR1), which affects tropoelastin secretion and cell survival. ARCL type III is also called Barsy syndrome. It is characterized by ophthalmic involvement (cataracts) and a reduction or absence of the corpus callosum that connects the two hemispheres of the brain. This type III is represented by mutations in the PYCR1 and ALDH18A1 genes.

Cutis laxa also has an X-linked form at the level of the ATP7A gene encoding a copper transporter across cell membranes, which will be detailed below in Menkes syndrome.

Acquired forms of cutis laxa are most often the consequence of generalized inflammatory events leading to the loss of elastic fibers by elastolysis. For example, penicillamine-based medication (lysyl oxidase inhibitor) can generate acquired cutis laxa.

Clinical detection of cutis laxa in children usually starts with a first observation of skin laxity, since this is what gives the syndrome its generic name. The skin is loose, inelastic and wrinkled. Research is refined toward the determination of a respiratory difficulty, inguinal hernias and/or joint laxity with, in particular, the protrusion of hip(s). Musculoskeletal developmental disorders may be associated: fractures, facial features, enlarged fontanelles, osteoporosis, scoliosis, joint dislocation, flat feet and low muscle tone. Alopecia (hair loss) may be observed. In addition to respiratory difficulties, cardiovascular and gastrointestinal problems, arterial tortuosity and urinary difficulties may occur. Developmental delays and cognitive difficulties may be added to the clinical picture, sometimes with epileptic seizures, microcephaly or macrocephaly. At the sensory level, myopia, ocular disorders and cataracts can be identified. No systemic therapy is known to date, and management is multidisciplinary, depending on the progress of the syndromes.

Supravalvular aortic stenosis (SVAS) and Williams-Beuren syndrome (WBS) lead to vessel obstruction. They can result from the complete excision of one copy of the elastin gene and its chromosomal environment (WBS) or from the production by one of the two alleles, as a result of a mutation, of an unstable product or a non-functional protein (SVAS). It should be noted that mice without elastin die 4 days after birth from total obstruction of the large vessels, due to uncontrolled smooth muscle cell proliferation. Indeed, a decrease in elastin synthesis at the prenatal stage results in an increase in cell proliferation and paradoxically in the number of elastic laminae in the aortic wall (Dietz and Mecham 2000). Williams-Beuren syndrome is related to a chromosomal microdeletion encompassing 26–28 genes including the elastin gene. The children have a characteristic physiognomy which led to the name of elf syndrome. Neuronal and cognitive development is affected, with abnormalities of connective and elastic tissue.

A.6.2. *Menkes syndrome and defects in copper metabolism*

Menkes disease or Wilson's disease are linked recessive diseases that lead to early death of the patients. The cause is related to a disorder of copper metabolism, characterized primarily by progressive neurodegeneration and marked connective tissue abnormalities (Scheiber et al. 2013). They are characterized by structural defects of connective tissue, mental disability as well as skeletal abnormalities including osteoporosis. These diseases are manifested by a hyper elasticity of the skin and severe skeletal abnormalities. Most children have stunted growth, feeding difficulties, vomiting and diarrhea. Their hair is depigmented and wire-like. There are several forms, but the course is usually severe.

The gene affected in Menkes disease is the ATP7A gene, and it is the ATP7B gene in Wilson disease. The expression of ATP7A is ubiquitous except for the liver and brain, whereas ATP7B is specific to the liver. These membrane proteins function as cation pumps involved in copper transport.

These mutations induce copper accumulation in certain tissues and dysfunction of lysyl oxidases, and thus the formation of covalent elastin bridges.

Cutis laxa linked to the X chromosome, recessive, is also called occipital horn syndrome or Ehlers-Danlos syndrome type IX. It is allelic with Menkes syndrome. Affected patients suffer from numerous connective tissue disorders (loose skin, hyper-extensibility of loose skin, hyper extensibility of the joints, bladder diverticulum) and sometimes neurological problems.

De Barsy-Moens-Dierckx syndrome, an acquired or genetic form, has also been linked to a significant decrease in elastic fibers without the genetic origin being known.

A.6.3. *Marfan syndrome and microfibril formation defects*

Marfan syndrome (MFS) is a systemic connective tissue disease characterized by a variable combination of cardiovascular, musculoskeletal, ophthalmic and pulmonary manifestations. It is characterized by excess growth of the long bones, ectopia lentis (dislocation or displacement of the natural crystalline lens) and, above all, aortic dissection with its risk of rupture as well as mitral insufficiency, which can be complicated (arrhythmias, endocarditis, heart failure). Ocular complications can lead to blindness. The fibrillin-1 gene, which forms the microfibrils that form the basis of elastic fibers, is mutated in the vast majority of cases. More than 500 mutations have been identified in the fibrillin-1 gene; they are responsible for MFS or various fibrillinopathies whose symptomatic characteristics form a continuum with those of MFS. For example, Shprintzen-Goldberg syndrome is a rare disease with features of MFS including facial dysmorphism, neonatal hypotonia and mental disability. A mutation in exon 29 of the FBN1 gene has been identified. Systemic sclerosis or human scleroderma is also associated with fibrillin-1, with cutaneous and visceral fibrosis and vascular lesions. Abdominal aortic aneurysms have also been associated with point mutations in the FBN1 gene, at exons 26 and 27. The transmission of the trait is autosomal dominant. A few sporadic cases have been reported. Management is multidisciplinary depending on the risks involved. Other more minority genes have been reported such as TGFB2, TGFB3, TGFBR1, TGFBR2 and SMAD3, all encoding elements of the TGF-beta pathway of which fibrillin is a major regulator (Coelho and Almeida 2020).

A.6.4. *Pseudoxanthoma elasticum and ectopic calcification of elastic tissue*

Pseudoxanthoma elasticum (PXE) is characterized by the progressive calcification and fragmentation of elastic fibers. It is a rare systemic metabolic disorder, recognized by its skin lesions, with damage to the connective tissue and the elastic fibers of the skin, the retina (and Bruch's membrane), as well as the arterial walls. Calcification of other organs may be observed, such as the heart, kidneys, breasts, pancreas, testicles, liver and spleen.

Mutations in the ABCC6 gene are responsible for PXE (Verschuere et al. 2021). This gene codes for the MRP6 (multidrug resistance associated protein) protein, which belongs to the ABC (ATP-binding cassette) family of membrane transporters. Its biological function is not clear, but it plays a role in cellular detoxification. The mutated ABCC6 and ENPP1 genes are likely to be part of a metabolic pathway leading to a decrease in circulating pyrophosphate, a major factor responsible for preventing calcification. Recently, mutations in the GGCX gene have been associated with a skin PXE phenotype and vitamin K-dependent coagulation factor deficiency. GGCX encodes an enzyme that catalyzes the carboxylation of coagulation factors, as well as the matrix protein Gla (MGP). This protein inhibits pathological mineralization of tissues, especially vessels, when activated by carboxylation. The current hypothesis is that the mutation on ABCC6 results in reduced gamma-glutamyl carboxylation of MGP, with subsequent mineralization (Boraldi et al. 2013).

A.6.5. *Other syndromes associated with elastic capital deficits*

Costello and Hurler syndromes are related to mutations in lysomal alpha-L-iduronidase. The symptoms seem to be due to the deficiency in the functioning of the elastin-binding protein, linked to an accumulation of chondroitin sulfate and dermatan sulfate (Hinek et al. 2004).

Certain types of glaucoma are part of fiber regulation. The discovery of polymorphisms in the LOXL1 gene as the only currently known risk factor for glaucoma (by pseudoexfoliation, PEX) was recently relayed by the demonstration of aberrant LOXL1 expression (Schlötzer-Schrehardt et al. 2008). LOXL1 is found in PEX pathological aggregates at the initial stages of elastogenesis and at different intra- and extra-ocular levels.

Ehlers-Danlos syndromes concern abnormalities of collagen fibers. They have a great genetic heterogeneity, with mainly cutaneous and articular hyperlaxity. The type defined as classical is related to a type V collagen defect, the vascular type is related to a type III collagen defect, while the genetic origin of the so-called "hypermobile" type is unknown.

Osteogenesis imperfecta is characterized by severe bone fragility due to mutations mainly in the COL1A1 and COL1A2 genes coding for the alpha1 and alpha2 chains of type I collagen. Other rarer but severe forms exist, with mutations in genes coding for proteins involved in the formation of collagen fibers.

A.7. Aging and elastic fibers

Intrinsic aging is an inevitable and irreversible physiological process due essentially to cellular aging. Elastic fibers have a very long life span and can therefore be present throughout life, but their low renewal rate makes their preservation particularly important during aging. Unfortunately, they are subject to multiple aggressions that alter or even annihilate their function and make them very limiting, thus becoming essential factors of longevity (Robert and Labat-Robert 2015). This is particularly true for the skin, with a marked disintegration of the elastic fibers of the papillary dermis, especially under the dermal-epidermal junction. This degradation concerns the aesthetic level but also the functional level as during the formation of hernias, phenomenon associated with the loss of elastic fibers at the level of the peritoneum (Pascual et al. 2009). This is also observed in vessels during the formation of varicose veins (Pascual et al. 2008) or the alteration of ligaments intervertebral discs (Yu 2002) or any other part of the body where they are present.

Factors contributing to aging of the skin and tissues are intrinsic and extrinsic and given as follows.

A.7.1. Glycation or non-enzymatic glycosylation

This so-called Maillard reaction is associated with the aging of tissues in general and elastic fibers in particular. It consists of the reaction of free amino groups of a lysine, hydroxylysine or arginine residue with reducing sugars. It leads to the formation of intra- and inter-molecular bridges, such as pentosidine, pyrraline and carboxymethyl-lysine, which reduce the physico-chemical capacities of elastic fibers. The products formed (advanced glycation end products [AGEs]) increase with aging and particularly with diabetes.

A.7.2. Modifications of already formed elastic fibers

Elastin also has a high propensity to retain lipids, which accelerates its degradation by proteases and reduces its elasticity (Jacob 2006). Between 20% and 30% of the weight of the "elastin" extracted from atherosclerotic plaques are cholesterol esters that are intimately linked to it. Calcium also has a strong affinity for elastin and potentiates the binding of cholesterol to elastin. The participation of proteoglycans in the structure of elastic fibers evolves with aging, which can weaken them (Gheduzzi et al. 2005).

A.7.3. The degradation of elastic fibers

Enzymes that break down proteins (proteases) can attack elastic fibers. There are several types of these elastases: the main group are serine elastases including pancreatic elastase, neutrophil elastase or leukocyte elastase and cathepsin G; serine elastases that are generally leukocyte and pancreatic; metal elastases including matrix metalloproteinases and finally cysteine elastases including some cathepsins whose enzymatic activity depends on a relatively acidic pH (5.5–6.5). Most of these proteases are associated with increased inflammation, whether post-traumatic, infectious or simply associated with aging or to a so-called "inflammatory" diet.

Elastin degradation leads to the release of soluble elastic peptides that can amplify an inflammatory state (Debret et al. 2005). They are notably found in the serum of patients with hypercholesterolemia, diabetes or hypertension. These peptides can activate numerous cellular signals and stimulate an inflammatory vicious circle leading to the creation of an inflammatory microenvironment and thus participating in the attraction of inflammation cells (leukocytes and macrophages) to the site. They also induce an increase in the expression of elastases (Wahart et al. 2019). Elastin peptides induce an acidification of the microenvironment that will promote the activity of cysteine elastases thus allowing the amplification of the degradation phenomenon. This contributes to the concentration of inflammation molecules in the matrix and are a form of immune defense set up against aging (Antonicelli et al. 2007).

A.7.4. The alteration of fibers by oxidation and photoaging

The accumulation of reactive oxygen species (ROS) is involved in the aging of elastic tissues (Figure 5.2). This is the case for components emitted by objects under combustion (cigarettes) or by industrial nanoparticles. Ultraviolet (UV) rays also induce ROS, which will alter the DNA and also the elastin. Paradoxically, ROS stimulate the synthesis of a poorly constituted and non-functional elastin that will contribute to the formation of hard tissue. Thus, UV rays induce this solar elastosis that results in a massive accumulation of non-elastic material in the upper and middle dermis.

Photoaging of the skin is a slow process, requiring several years before becoming clinically apparent. It is also manifested by the appearance of wrinkles, dyschromias, telangiectasias, pre-epitheliomatous lesions, xerosis and cutaneous fragility. Histologically, photoaging is mainly marked by dermal elastosis. Physiopathologically, these phenomena result from the cumulative and synergistic effects of UVB, UVA and infrared rays, which promote the formation of free radicals. The aging tissue results in atrophy of the dermis, a decrease in the metabolism of fibroblasts, which become senescent, and thus a decrease in the synthesis of the extracellular matrix. There is an increase in proteinase activity leading to an accumulation of fragmented collagen, an increase in cross-linking and glycosylation of collagen. This gives collagen greater resistance to digestion by collagenases. On the other hand, we observe an alteration of the structure of the elastic fibers by the presence of calcium deposits as well as by an increase in their degradation by elastases. This aging is accompanied by a reduction in the quantity of glycoaminoglycans (hyaluronic acid and dermatan sulphate in particular) leading to the alteration of the viscoelastic and hydration properties of the skin.

A.8. Elasticity and its maintenance

A.8.1. *Pharmacology in a few examples*

Although it is not possible to replace, stimulate and regenerate elastic tissues in their naturally completed form, several therapeutic approaches are proposed with interesting results. Some direct or indirect targeting can be stated as follows.

(1) Indirect targeting

– Cardiovascular and renal regulation: targeting hypertension via the renin–angiotensin–aldosterone system regulation of vascular tonicity (protein, sugar and blood salt levels) and heart rhythm (arrhythmia, fibrillation);

– respiratory failure: bronchodilation by reversing the constrictor effect of the parasympathetic system by the vagus nerve (anticholinergics) or mimicking the vasodilator system of the sympathetic system (beta-2 agonists);

– regulation of diabetes: insulin, glucose regulation (IGF1 pathway);

– inflammation: anti-inflammatory (osteoarthritis, fibrosis retardant);

– hypodigestion, diverticulitis: regulation of the food bolus;

– hyper-calcification and hyper-lipidation/decalcification: vitamin D => diet, statins.

(2) Direct targeting

– Activation of collagen synthesis: vitamin C, copper, vitamin D, folate, vitamin B12 and B6;

– activation of the synthesis of functional elastic fibers: dill seed extract (unknown active ingredient);

– reduction of pathological cross-linking: cross-linking inhibitor (D-penicillamine), non-enzymatic glycation inhibitor (advance glycation end products – breakers <> related to excess sugar).

A.8.2. *The role of food and intakes*

Since connective tissue fibers are made up of proteins, a diet providing the amino acids that constitute them can be beneficial. In the same way, the stimulation of their synthesis and their assembly depends on several factors, which must be provided by the diet. In addition, certain substances must be avoided. We can build a list of positive and negative intakes.

(1) Positive intakes (sufficient and not excessive)

– The major amino acids in fibers are glycine (white meat, fish, vegetables), valine (lean meat, eggs, fish, vegetables), proline (soy protein, eggs, cheese, lean meat, fish, cereals, fruits and vegetables) and glutamine (eggs, dairy products, cabbage, parsley);

– minerals essential to the formation of fibers: copper, iron, magnesium, zinc, manganese, sulfur or sulfides and calcium;

– essential vitamins: vitamins C, D, B6 and B12 and anti-oxidants (vitamins E, C, D, flavonoids and beta-carotene);

– plant extracts to build and/or protect, such as dill, fennel, and plants with a proven anti-oxidant effect.

(2) Negative intakes to be avoided (in absolute terms or if ingested in excess)

– Excessive sugars, to avoid glycation of proteins and formation of AGE;

– excessive red meat, to avoid folate, vitamin B12 and B6 deficiency, with the potential consequence of homocysteinemia;

– fatty acids in excess, to avoid the degradation of the elastic walls of the arteries and the formation of atheromatous plaques => aneurysm;

– processed and oxidizing foods in excess, such as hydrogenated oils, alcohol, food additives, nitrogenous residues from intensive agriculture and smoked products;

– microparticles (from combustion or degradation of plastics);

– pesticides (known or potential) such as succinate dehydrogenase inhibitors or glyphosate to avoid potential inhibition of mitochondria and the formation of elastic and collagenous fibers;

– volatile fibers (asbestos/metal residues and construction products), in order to avoid respiratory ailments such as sarcoidosis, chronic obstructive pulmonary disease, bronchitis and even mesothelioma;

– infectious agents, of viral, bacterial and parasitic origin, by favoring preventive (protective gestures) or therapeutic (vaccination when available) precautions.

A.9. Elasticity and movement

To date, the best way to maintain the elasticity of the body is through movement. Multiple approaches can be recommended depending on the intentions or situations. Some of the approaches mentioned in the text can be briefly listed, such as:

– gentle strengthening and flexibility: synthesis of fibers and muscles (fight against sarcopenia), ad hoc calcification and stretching of elastic proteins of muscles (titin) and fibers (fascias, ligaments, and tendons);

– the fight against osteoporosis and muscle wasting, stimulation to avoid the loss of cushioning (viscoelastic cartilage);

– the mobilization of the six senses, the sensorimotor areas and the central and peripheral nervous systems;

– stimulation of respiration (increase in mitochondria);

– stimulation of digestion and glycemic regulation;

– stimulation of oxygen fixation;

– strengthening of the lymphatic sphincters (massage);

– avoidance of chronic trauma: well-balanced walking, identification of elastic limits in repetitive motions and hydration;

– training blood pressure by alternating hot and cold (contraction and dilation of the veins);

– exercises in restraint clothing if necessary;

– birth preparation: vaginal stretching and perineum preparation;

– balancing the body's electrical gradient with regular grounding (earthing or grounding).

A.10. The laws of breathing and yoga

– Slow breathing in for efficient gas entry and dissolution. O_2 spontaneously enters the body from the air (where it is more concentrated) to the body (where it is less concentrated). It is the opposite for CO_2 produced in the body by the combustion of oxygen (and sugars). CO_2 will therefore tend to escape to the air, if and only if this space does not contain a higher concentration of CO_2.

– Inspiration by extending the diaphragm downward, which inflates the lower elastic parts of the thorax (and not the upper, more rigid rib cage). The extension accelerates the diffusion of oxygen by generating a kind of air draft by increasing the volume of the lungs. The gases diffuse under the sub-alveolar blood vessels. Oxygen is taken up by hemoglobin.

– Retention of respiration after inspiration for a facilitated competition between O_2 and CO_2 for their carrier, hemoglobin, and for the cells of the tissues, thus increasing the energy efficiency of respiration. CO_2 drives oxygen away from hemoglobin, which will reach the cells more efficiently.

– Exhalation (passive decrease in lung volume) increases the concentration of CO_2 in the blood. CO_2 can escape more spontaneously into the air, where it is less concentrated, as it passes through the veins below the lungs.

– Circulation allows O_2 and CO_2 carried by the hemoglobin in the blood to flow from the blood vessels within the lungs to the tissues that need it. This is made possible by the synchronized action of the two parts of the heart.

A.11. Epigenetic regulation

A.11.1. *From epigenetics to aging...*

The idea that all our cells have the same genome is incorrect. Mutations, deletions, duplications and transpositions of chromosomal fragments occur throughout life, either very locally at the cell level, in localized areas, or in mosaics. These genomic perturbations lead to gains or losses of function in the affected cells, which may eventually result in a pathological state. However, these genomic modulations are not sufficient to explain the fact that our organism is composed of more than 200 distinct cell types with different morphologies and functions. This particularity is directly linked to epigenetic mechanisms that allow the establishment and maintenance of an expression profile of genes specific to the cell type considered (Laurent et al. 2010).

As mentioned earlier in this book, epigenetics can be considered an overlay of the genetic code and has two main roles. The first is to maintain the stability of the genetic information by preventing, for example, the transposition of DNA fragments to an inappropriate location; this mainly concerns non-coding regions of the DNA (outside the genes). The second and most well-known role of epigenetics is to modulate gene expression. Thus, the character (we call it the phenotype) of a cell is dependent on the genome and the epigenome. In terms of gene expression control, there are several major groups of epigenetic players. The first two affect DNA or proteins that act on the compaction of DNA chemically and therefore also affect chromosomes. This compaction has an obvious mechanical role on the reading of genes on the chromosomes in this fundamental process called gene transcription. It is the reading of the DNA sequences into these coding intermediates that are the so-called "messenger" RNAs, which will serve as the basis of information to build proteins. Among these are the histones involved in the compaction of the DNA, which we mentioned above. The third group is made up of a whole range of non-coding RNAs (i.e. they are not used as a matrix for the shaping of proteins). These RNAs, which vary greatly in size, play an additional role in regulating protein formation, adding another level of complexity.

Among the recommendations of a symposium on aging led by Alfonso J. Cruz-Jentoft, Alain Franco, Andrzej Milewicz and Pascal Sommer, the issue of focusing research on epigenetics in aging was proposed as a priority. The proceedings of the symposium supported by the European Community and organized in Wroclaw (Poland) from September 11 to 13, 2008, were reported in the European silver paper (on the future of health promotion and preventive actions, basic research and clinical aspects of age-related diseases), which was widely disseminated in many scientific journals in the field (Cruz-Jentoft et al. 2009).

Although all epigenetic marks are of equal importance in terms of molecular mechanisms, DNA methylation is probably the epigenetic mark that has been most studied in aging. Methylation is a simple, targeted and reversible chemical modification of DNA. It is carried out by an enzyme whose name describes its function, since it is called DNA methyltransferase (we mention it because it will play an important role in the rest of the presentation). The corresponding epigenetic imprint can be added during cell multiplication, during DNA repair (because DNA is highly monitored), or in response to signals from the cell's environment. It is therefore a dynamic mechanism, yet certain specific regions of the DNA of the chromosomes will progressively remain methylated over time to constitute sorts of aging zones with a high methylation rate (the aging-associated differentially methylated regions). We can thus observe dense or non-dense areas of methylation clusters on the DNA. The dense regions are mostly found in the very active (and coding) regions of the DNA, while the sparse regions are rather represented in the non-coding intergenic sequences. While the sparse regions essentially control chromosomal stability, the dense islands are directly responsible for gene expression, and their methylation status predicts whether the gene is activatable or not in the cell in question, given that DNA methylation is generally associated with gene inactivation. This repressed status can be found in many biological phenomena such as the inactivation of the extra X chromosome in females, the decreased expression of genes with parental imprinting, or during aging. As a striking example of the role of methylation in aging, in 2014 Wolfgang Wagner's group was able to demonstrate using blood samples that observing the methylation state of only three specific sites in the DNA allowed the chronological age of an individual to be traced to within 5 years (these are in fact three so-called CpG clusters located in the ITGA2B, ASPA and PDE4C genes; Weidner et al. 2014).

A.11.2. ...towards elastic fibers

The gene coding for the various components of elastic fibers are not immune to epigenetic control and its inexorable drift during aging. In this book, we have discussed the case of an elastin cross-linking enzyme, LOXL1. It turns out that the LOXL1 gene also undergoes hypermethylation of its proximal promoter in a region rich in these CpG dinucleotides (Figure 5.1). This region controls the reading of the LOXL1 gene. The methylations contributed by one of these DNA methyltransferases previously introduced prevent the reading of this gene by the cell's machinery. These methylations are more abundant in elderly subjects, which contributes to a decrease in LOXL1 synthesis and thus to a decrease in the formation of elastic fibers in the extracellular space. It could therefore be deduced from this finding that blocking the methylation process of these control areas of the LOXL1 gene could restart a satisfactory synthesis of fibers. And indeed, treatment of cells with a DNA methyltransferase inhibitor (5-aza-2'-deoxycytidine for specialists) can activate a more consistent synthesis of elastic fibers by reducing hypermethylation of the LOXL1 promoter. However, this pharmacological compound used in chemotherapy is highly cytotoxic and may induce demethylation of genes coding for anti-oncogenes, especially during long-term use.

In an approach similar to that used to identify dill extract, we looked for non-toxic active ingredients targeting the activity of a particular DNA methyltransferase, called isoform 3A (or "DNMT3A") in a more moderate way than the drug used previously. We were thus able to select an extract of marjoram (Moulin et al. 2017) among other ingredients that showed some efficacy on cells in culture and on reconstructed skin. To go further in the search for a pharmacological agent, it was important to dig deeper into the molecular mechanism of LOXL1 hypermethylation, which is not limited to DNMT3A activity or gene expression. We were thus able to discover two molecular partners that participate in the action of DNMT3A associated with aging, making LOXL1 (and elastic fiber) synthesis less efficient (Figure 5.1). These partners are protein arginine methyltransferase 5 (PRMT5) and methylosome protein 50 (MEP50). In young individuals, acetylation of histones (green circles) by histone acetyltransferases promotes a relaxed chromatin structure and therefore permissive for transcription. In older individuals, histone deacetylation occurs, and PRMT5 adds methyl groups to the C-terminal tails of histones. This promotes the recruitment of DNMT3A, which interacts with histone and particularly di-methylated histone H4. In addition, the approach of nucleosomes compacts the DNA which increases the probability of DNMT3A encountering nearby CpG sites to methylate the DNA (gray circles). This creates a more stable transcriptionally inactive chromatin, and LOXL1 is less expressed. In the presence of active ingredients, it appears that the DNMT3A/PRMT5/MEP50 complex is destabilized preventing DNMT3A activity on DNA. This mechanistic hypothesis is further supported by the fact that plant extracts have no effect on the expression of the genes encoding these different partners.

The elastin gene does not seem to be affected by methylation modulation during aging, as elastin gene expression remains relatively constant after growth. However, another epigenetic mechanism is at work which again reflects the extraordinary complexity of life. At this stage, small non-coding

RNAs are involved. Specific microRNAs (miR-29) have been widely described as regulators of the stability of "messenger" RNAs coding for elastin, fibrillins and certain collagens. Thus, the presence of miR-29 in the cell will decrease the synthesis of these proteins, which may be responsible for a depletion of the extracellular matrix; or conversely, a decrease in these microRNAs authorizes an exacerbated secretion of elastin and collagens, as has been demonstrated in myocardial fibrosis following an infarction, for example (van Rooij et al. 2008). Although studies report modulations of miR-29 expression during aging, no link with elastic fibers has been put forward so far, but this mechanism of post-transcriptional regulation would deserve more attention to provide new means to fight pulmonary fibrosis following chronic obstructive pulmonary disease or after severe SARS-Cov2 infection.

A.11.3. *When elasticity controls epigenetics*

We have seen that elastic fibers can be regulated at the level of their genes or their "messengers" by epigenetic mechanisms, but can the reciprocal be true? Finally, could elasticity at the level of tissues and organs in turn control epigenetic mechanisms inside the cell? This question finds its legitimacy through the remarkable work initiated in the mid-2000s by Adam Engler and Dennis Discher (University of Pennsylvania), showing that by simply modifying a cell culture support's modulus of elasticity, stem cells spontaneously opted for a different destiny, becoming a neuronal, muscle or bone cell (Engler et al. 2006). This is a mechano-transduction mechanism, i.e., the transformation of a physical phenomenon into a biological mechanism in the cell, which leads to an epigenetic rearrangement of reprogramming. Until today, mechano-biological studies have gained interest in the scientific community by associating physicists and biologists, but also chemists to develop new biomaterials in regenerative medicine.

If we return to the question of LOXL1 methylation in aged cells, could this not be the consequence rather of the loss of elasticity? On closer inspection, our mechanistic discovery was first made using dermal fibroblasts from a young cutis laxa patient aged 8, whose etiology is based on a mutation in the gene coding for fibulin-5. At this age, aging cannot be the cause of such an epigenetic drift, but the fibulin-5 mutation drastically alters the synthesis and function of elastic fibers. While it is difficult to see a direct relationship between the mutation and the epigenetic mechanism, the fact that the tissues of this young patient lack elasticity allows a much more relevant parallel with the equivalent disorder found in elderly subjects. This hypothesis has very recently been echoed in studies by Daniel Greif and Jui Dave (Yale University, New Haven) on SVAS found in patients with Williams-Beuren syndrome and other genetic elastic deficiency syndromes. It appears that at the vascular level, the lack of elastin leads to uncontrolled proliferation of vascular smooth muscle cells, leading to vascular stenosis, by means of hypomethylation of a specific gene (and thus overactivation of the Notch pathway essential for regulating cell proliferation; Dave et al. 2022).

These recent studies show that there is a vicious circle between loss of elasticity – whether syndromic, pathological or due to chronological aging – and epigenetic reprogramming. Treatments targeting the epigenetic mechanisms themselves are complex to implement because they are rarely specific to a particular gene and risk inducing more or less deleterious side effects; nevertheless, these strategies have the merit of existing and being developed. Another strategy to stop this vicious circle could be to repair the elastic fibers themselves. Given the complexity of the structure of elastic fibers, it is difficult to conceive of reconstructing these fibers identically. However, a bit like an orthopedic surgeon who builds a prosthesis to replace a joint and restore its function, even if it is diminished, providing a molecular entity capable of reinforcing the elasticity of a biological tissue seems possible today because of the progress of genetic engineering and biotechnologies. This "elastin-drug" is not yet available, but who knows, the near future may surprise us.

A.12. Elasticity and bioengineering

A great number of materials have been designed to remedy a defect in the body's tissues. The implantation of prostheses made of titanium, ceramics or calcium derivatives as well as dental composites, medical textiles or intravascular mesh (stents) has become a routine matter. Their obvious quality is that they meet a clear functional need and their success is considerable, and rightly so. Research continues to progress with a major trend to offer more biocompatibility through a biomimetic approach. While success concerns mainly "hard" tissues (bones, teeth), for "soft" elastic tissues, the challenge is to develop materials that will either be improved to integrate as well as possible in an environment in perpetual deformation, or composed in such a way as to generate an elastic biomaterial with the characteristics of the one to be replaced. To date, three approaches have been favored: the synthesis of elastic materials with no real similarities to existing elastic fibers, the use of a part of tropoelastin to elaborate a biomimetic elastic structure and the formation of a composite structure combining these parts. This is the third solution that was developed in the Laboratory of Tissue Biology and Therapeutic Engineering under the leadership of Drs. Debret and Sohier. They first developed an elastic protein with the same physicochemical characteristics as human elastin and capable of integrating into living tissue. Then, they associated this protein with a modular gel compatible with living elastic tissues. This elastin-mimetic gel continues to be developed for various applications in tissue engineering. In particular, it is possible to add woven microfibers to resist more strongly the strains observed in vessels. By changing the formulations of the composite material, it is then possible to mold the whole to mimic the behavior of a vein or an artery. This device had been supported by the CNRS in order to manufacture artificial vessels for training in microsurgery. The association of the gel with microfibers is also the strategy that has been chosen in the framework of another project of reconstruction of vocal cords with laboratories in Grenoble specialized in mechanics and phonation. Although implantation in humans is still quite far off, certain formulations make it possible to obtain full-scale objects capable of self-oscillating to produce a sound – the gel sings! Another challenge has been to make the gel porous after injection in the context of muscle and heart reconstruction. It is impressive to see the cardiac cells (cardiomyocytes) organize themselves and make the structure beat like a mini heart. Finally, the synthetic elastic protein used alone is also of growing interest with the development of biotherapies. This protein is currently being tested in an arterial elasticity repair program (www.arterylastic.com) and in skin substitutes.

The identification of the elastic characteristics of living tissues has greatly improved the performance of engineering for health, whether in terms of advances in medical imaging, indentation hardness and tribology, or the development of customized prostheses and orthotics. Elastography techniques have become routine in medical imaging. Very schematically, they are based on the application of an electromagnetic or ultrasonic signal whose complex path is analyzed and mathematically processed to build a qualitative or even quantitative imaging. However, the evaluation of mechanical properties of tissues remains complex because it is generally limited to the evaluation of a single parameter at a time, by applying a deformation, compression or shear, for example. Different methods co-exist depending on the definition required, from the use of ultrasound in standard elastography to dynamic magnetic resonance.

Ultrasound techniques applied to medicine have been developed since the 1950s. They derive from the development of sonar. These echographic techniques (by analysis of the echo, i.e. the reflected diffusion of sounds) are non-invasive and non-destructive for the tissues, and they allow a rapid follow-up of organs by combining rapid echography, the Doppler effect and the recording of movements. The resolution can be improved by elastography, which adds the elastic parameters of the tissue to the ultrasound information. Dynamic elastography (or sonoelastography) adds to this the internal generation of sound waves by the tissues under stress. All in all, it is a question of analyzing the complex interplay of reflected waves that form a kind of wave speckle (the "acoustic speckle") that the calculation makes it possible to differentiate into graphic information and parameterization on the elasticity. Coupling with other imaging modalities can further increase the resolution and the amount of information that can be identified in the tissues, taking into account the mechanical properties of the tissues without destroying them (Kammoun et al. 2019). This is the case, for example, with magnetic resonance elastography, which allows the study of the viscoelastic properties of biological tissues by visualizing the propagation of shear waves in a tissue subjected to low frequency vibration.

References

Alegre-Cebollada, J. (2021). Protein nanomechanics in biological context. *Biophysical Reviews*, 13(4), 435–454.

Antonicelli, F., Bellon, G., Debelle, L., Hornebeck, W. (2007). Elastin-elastases and inflamm-aging. *Current Topics in Developmental Biology*, 79, 99–155.

Bäck, M., Aranyi, T., Cancela, M.L., Carracedo, M., Conceição, N., Leftheriotis, G., Macrae, V., Martin, L., Nitschke, Y., Pasch, A. et al. (2018). Endogenous calcification inhibitors in the prevention of vascular calcification: A consensus statement from the COST Action EuroSoftCalcNet. *Frontiers in Cardiovascular Medicine*, 5, 196.

Baldwin, A.K., Simpson, A., Steer, R., Cain, S.A., Kielty, C.M. (2013). Elastic fibres in health and disease. *Expert Reviews in Molecular Medicine*, 15, e8.

Berthoz, A. (2013). *Le sens du mouvement*. Odile Jacob, Paris.

Berthoz, A. (2020). *L'inhibition créatrice*. Odile Jacob, Paris.

Beyens, A., Pottie, L., Sips, P., Callewaert, B. (2021). Clinical and molecular delineation of cutis laxa syndromes: Paradigms for homeostasis. *Advances in Experimental Medicine and Biology*, 1348, 273–309.

Bhatti, J.S., Bhatti, G.K., Reddy, P.H. (2017). Mitochondrial dysfunction and oxidative stress in metabolic disorders – A step towards mitochondria based therapeutic strategies. *Biochimica et Biophysica Acta. Molecular Basis of Disease*, 1863(5), 1066–1077.

Bianchi-Demicheli, F. and De Ziegler, D. (2005). Pharmacological treatment of female sexual disfunction: Chimera or reality? *Revue médicale suisse*, 1(11), 749–753.

Blackburn, E.H., Epel, E., Wiart, Y. (2017). *L'effet télomère : une approche révolutionnaire pour allonger sa vie et ralentir les effets du vieillissement.* Guy Tredaniel éditeur, Paris.

Boizot, J., Minville-Walz, M., Reinhardt, D.P., Bouschbacher, M., Sommer, P., Sigaudo-Roussel, D., Debret, R. (2022). FBN2 silencing recapitulates hypoxic conditions and induces elastic fiber impairment in human dermal fibroblasts. *International Journal of Molecular Sciences*, 23(3), 1824.

Boraldi, F., Annovi, G., Vermeer, C., Schurgers, L.J., Trenti, T., Tiozzo, R., Guerra, D., Quaglino, D. (2013). Matrix gla protein and alkaline phosphatase are differently modulated in human dermal fibroblasts from PXE patients and controls. *The Journal of Investigative Dermatology*, 133(4), 946–954.

Boraldi, F., Lofaro, F.D., Cossarizza, A., Quaglino, D. (2022). The "elastic perspective" of SARS-CoV-2 infection and the role of intrinsic and extrinsic factors. *International Journal of Molecular Sciences*, 23(3), 1559.

Bordoni, B. and Zanier, E. (2014). Clinical and symptomatological reflections: The fascial system. *Journal of Multidisciplinary Healthcare*, 7, 401.

Boucand, M.-H. (2018). *Une approche éthique des maladies rares génétiques : enjeux de reconnaissance et de compétence.* Éditions Érès, Toulouse.

Brillat-Savarin, J.A. (1981). *Physiologie du goût.* Édition mise en ordre et annotée. Hermann, Paris.

Bronner, G. (2021). *Apocalypse cognitive.* Presses universitaires de France, Paris.

Campagnac-Morette, C. (2021). *Prévenir & guérir par le yoga : exercices et postures*, New edition. Editions du Dauphin, Paris.

Carrancá, M., Griveau, L., Remoué, N., Lorion, C., Weiss, P., Orea, V., Sigaudo-Roussel, D., Faye, C., Ferri-Angulo, D., Debret, R. et al. (2021). Versatile lysine dendrigrafts and polyethylene glycol hydrogels with inherent biological properties: In vitro cell behavior modulation and in vivo biocompatibility. *Journal of Biomedical Materials Research. Part A*, 109(6), 926–937.

Castelucci, B.G., Consonni, S.R., Rosa, V.S., Sensiate, L.A., Delatti, P.C.R., Alvares, L.E., Joazeiro, P.P. (2018). Time-dependent regulation of morphological changes and cartilage differentiation markers in the mouse pubic symphysis during pregnancy and postpartum recovery. *PLoS One*, 13(4), e0195304.

Cenizo, V., André, V., Reymermier, C., Sommer, P., Damour, O., Perrier, E. (2006). LOXL as a target to increase the elastin content in adult skin: A dill extract induces the LOXL gene expression. *Experimental Dermatology*, 15(8), 574–581.

Charentenay, P.D. (2007). Les cinquante ans du Traité de Rome. *Etudes*, 406(3), 309–319.

Chastan, N. and Decker, L.M. (2019). Posturo-locomotor markers of preclinical Parkinson's disease. *Clinical Neurophysiology*, 49(2), 173–180.

Chevalier, G., Sinatra, S.T., Oschman, J.L., Sokal, K., Sokal, P. (2012). Earthing: Health implications of reconnecting the human body to the earth's surface electrons. *Journal of Environmental and Public Health*, 2012, 291541.

Chiousse, S. (1995). Divins thérapeutes – La santé au Brésil revue et corrigée par les orixas. Anthropologie sociale et ethnologie. Ecole des Hautes Etudes en Sciences Sociales (EHESS). Open HAL ID: tel-00003395. Available at: https://tel.archives-ouvertes.fr/tel-00003395 [Accessed 12/06/2021].

Claus, S., Fischer, J., Mégarbané, H., Mégarbané, A., Jobard, F., Debret, R., Peyrol, S., Saker, S., Devillers, M., Sommer, P., Damour, O. (2008). A p.C217R mutation in fibulin-5 from cutis laxa patients is associated with incomplete extracellular matrix formation in a skin equivalent model. *The Journal of Investigative Dermatology*, 128(6), 1442–1450.

Clausen, T.M., Sandoval, D.R., Spliid, C.B., Pihl, J., Perrett, H.R., Painter, C.D., Narayanan, A., Majowicz, S.A., Kwong, E.M., McVicar, R.N. et al. (2020). SARS-CoV-2 infection depends on cellular heparan sulfate and ACE2. *Cell*, 183(4), 1043–1057.e15.

Cocciolone, A.J., Hawes, J.Z., Staiculescu, M.C., Johnson, E.O., Murshed, M., Wagenseil, J.E. (2018). Elastin, arterial mechanics, and cardiovascular disease. *American Journal of Physiology: Heart and Circulatory Physiology*, 315(2), H189–H205.

Coelho, S.G. and Almeida, A.G. (2020). Marfan syndrome revisited: From genetics to the clinic. *Revista Portuguesa De Cardiologia*, 39(4), 215–226.

Craft, C.S., Pietka, T.A., Schappe, T., Coleman, T., Combs, M.D., Klein, S., Abumrad, N.A., Mecham, R.P. (2014). The extracellular matrix protein MAGP1 supports thermogenesis and protects against obesity and diabetes through regulation of TGF-β. *Diabetes*, 63(6), 1920–1932.

Cruz-Jentoft, A.J., Franco, A., Sommer, P., Baeyens, J.P., Jankowska, E., Maggi, A., Ponikowski, P., Rys, A., Szczerbinska, K., Michel, J.-P., Milewicz, A. (2009). Silver paper: The future of health promotion and preventive actions, basic research, and clinical aspects of age-related disease – A report of the European Summit on Age-Related Disease. *Aging Clinical and Experimental Research*, 21(6), 376–385.

Csepe, T.A., Kalyanasundaram, A., Hansen, B.J., Zhao, J., Fedorov, V.V. (2015). Fibrosis: A structural modulator of sinoatrial node physiology and dysfunction. *Frontiers in Physiology*, 6, 37.

Dave, J.M., Chakraborty, R., Ntokou, A., Saito, J., Saddouk, F.Z., Feng, Z., Misra, A., Tellides, G., Riemer, R.K., Urban, Z. et al. (2022). JAGGED1/NOTCH3 activation promotes aortic hypermuscularization and stenosis in elastin deficiency. *The Journal of Clinical Investigation*, 132(5), e142338.

Debret, R., Antonicelli, F., Theill, A., Hornebeck, W., Bernard, P., Guenounou, M., Le Naour, R. (2005). Elastin-derived peptides induce a T-helper type 1 polarization of human blood lymphocytes. *Arteriosclerosis, Thrombosis, and Vascular Biology*, 25(7), 1353–1358.

Debret, R., Cenizo, V., Aimond, G., André, V., Devillers, M., Rouvet, I., Mégarbané, A., Damour, O., Sommer, P. (2010). Epigenetic silencing of lysyl oxidase-like-1 through DNA hypermethylation in an autosomal recessive cutis laxa case. *The Journal of Investigative Dermatology*, 130(11), 2594–2601.

Decorps, J., Saumet, J.L., Sommer, P., Sigaudo-Roussel, D., Fromy, B. (2014). Effect of ageing on tactile transduction processes. *Ageing Research Reviews*, 13, 90–99.

Dietz, H.C. and Mecham, R.P. (2000). Mouse models of genetic diseases resulting from mutations in elastic fiber proteins. *Matrix Biology: Journal of the International Society for Matrix Biology*, 19(6), 481–488.

Dubnikov, T., Ben-Gedalya, T., Cohen, E. (2017). Protein quality control in health and disease. *Cold Spring Harbor Perspectives in Biology*, 9(3), a023523.

Dussart, C., Ngo, M.-A., Siranyan, V., Sommer, P. (2019). *De la démocratie sanitaire à la démocratie en santé : acte du colloque qui s'est tenu à la MSHS de Nice, le 3 décembre 2018*. LEH édition, Bordeaux.

Ekamper, P., van Poppel, F., Stein, A.D., Lumey, L.H. (2014). Independent and additive association of prenatal famine exposure and intermediary life conditions with adult mortality between age 18–63 years. *Social Science & Medicine*, 119, 232–239.

Ellul, J. (1990). *La technique : ou, L'enjeu du siècle*, 2nd edition. Revue Economica, Paris.

Ellul, J. and Rognon, F. (2008). *L'Apocalypse : architecture en mouvement*. Editions Labor et fides, Geneva.

Emmanuelli, X. and Cyrulnik, B. (2021). *Se reconstruire dans un monde meilleur*. Editions Humensciences, Paris.

Engler, A.J., Sen, S., Sweeney, H.L., Discher, D.E. (2006). Matrix elasticity directs stem cell lineage specification. *Cell*, 126(4), 677–689.

Ferry, L. and Ferry, L. (2013). *Une brève histoire de l'éthique*. Le Figaro le Point, Paris.

Fhayli, W., Boëté, Q., Kihal, N., Cenizo, V., Sommer, P., Boyle, W.A., Jacob, M.-P., Faury, G. (2020). Dill extract induces elastic fiber neosynthesis and functional improvement in the ascending aorta of aged mice with reversal of age-dependent cardiac hypertrophy and involvement of lysyl oxidase-like-1. *Biomolecules*, 10(2), E173.

Fournet, M., Bonté, F., Desmoulière, A. (2018). Glycation damage: A possible hub for major pathophysiological disorders and aging. *Aging and Disease*, 9(5), 880–900.

Gayon, J. and Petit, V. (2019). *Knowledge of Life Today: Conversations on Biology*. ISTE Ltd., London, and John Wiley & Sons, New York.

Gennisson, J.-L., Deffieux, T., Fink, M., Tanter, M. (2013). Ultrasound elastography: Principles and techniques. *Diagnostic and Interventional Imaging*, 94(5), 487–495.

Gigante, A., Chillemi, C., Quaglino, D., Miselli, M., Pasquali-Ronchetti, I. (2001). DL-penicillamine induced alteration of elastic fibers of periosteum-perichondrium and associated growth inhibition: An experimental study. *Journal of Orthopaedic Research: Official Publication of the Orthopaedic Research Society*, 19(3), 398–404.

Gluck, J.M., Herren, A.W., Yechikov, S., Kao, H.K.J., Khan, A., Phinney, B.S., Chiamvimonvat, N., Chan, J.W., Lieu, D.K. (2017). Biochemical and biomechanical properties of the pacemaking sinoatrial node extracellular matrix are distinct from contractile left ventricular matrix. *PLoS One*, 12(9), e0185125.

Gurevich, E.V. and Gurevich, V.V. (2015). Beyond traditional pharmacology: New tools and approaches: Expanding pharmacological toolbox. *British Journal of Pharmacology*, 172(13), 3229–3241.

Hinek, A., Braun, K.R., Liu, K., Wang, Y., Wight, T.N. (2004). Retrovirally mediated overexpression of versican v3 reverses impaired elastogenesis and heightened proliferation exhibited by fibroblasts from Costello syndrome and Hurler disease patients. *The American Journal of Pathology*, 164(1), 119–131.

Illich, I. (1973). *Tools for Conviviality*. Harper & Row, New York.

Illingworth, C.M. (1974). Trapped fingers and amputated finger tips in children. *Journal of Pediatric Surgery*, 9(6), 853–858.

Isselbacher, E.M. (2018). Losartan for the treatment of marfan syndrome. *Journal of the American College of Cardiology*, 72(14), 1619–1621.

Jacob, M.-P. (2006). Matrice extracellulaire et vieillissement vasculaire. *Médecine/sciences*, 22(3), 273–278.

de Jaeger, C., Fraoucene, N., Voronska, E., Cherin, P. (2010). Rôle de l'homocystéine en pathologie. *Médecine & longévité*, 2(2), 73–86.

Jeanblanc, A. (2014). 50 ANS DE L'INSERM. Pr Nicolas Lévy : de la progéria au vieillissement de la population. *Le Point* [Online]. Available at: https://www.lepoint.fr/editos-du-point/anne-jeanblanc/50-ans-de-l-inserm-pr-nicolas-levy-de-la-progeria-au-vieillissement-de-la-population-2014-1807465_57.php [Accessed 31 March 2014].

Joufflineau, C., Vincent, C., Bachrach, A. (2018). Synchronization, attention and transformation: Multidimensional exploration of the aesthetic experience of contemporary dance spectators. *Behavioral Sciences (Basel, Switzerland)*, 8(2), E24.

Kammoun, M., Ternifi, R., Dupres, V., Pouletaut, P., Même, S., Même, W., Szeremeta, F., Landoulsi, J., Constans, J.-M., Lafont, F. et al. (2019). Development of a novel multiphysical approach for the characterization of mechanical properties of musculotendinous tissues. *Scientific Reports*, 9(1), 7733.

Kell, D.B., Heyden, E.L., Pretorius, E. (2020). The biology of lactoferrin, an iron-binding protein that can help defend against viruses and bacteria. *Frontiers in Immunology*, 11, 1221.

Kenny, T.C., Gomez, M.L., Germain, D. (2019). Mitohormesis, UPRmt, and the complexity of mitochondrial DNA landscapes in cancer. *Cancer Research*, 79(24), 6057–6066.

Korff-Sausse, S. (2020). *Figures du handicap : mythes, arts, littérature*. Payot & Rivages, Paris.

Korgavkar, K. and Wang, F. (2015). Stretch marks during pregnancy: A review of topical prevention. *The British Journal of Dermatology*, 172(3), 606–615.

Kozel, B.A. and Mecham, R.P. (2019). Elastic fiber ultrastructure and assembly. *Matrix Biology: Journal of the International Society for Matrix Biology*, 84, 31–40.

Kropotkin, P.A. (2009). *L'entraide, un facteur de l'évolution*. Éditions du Sextant, Paris.

Kucenas, S. (2015). Perineurial glia. *Cold Spring Harbor Perspectives in Biology*, 7(6), a020511.

Kumar, A., Palfrey, H.A., Pathak, R., Kadowitz, P.J., Gettys, T.W., Murthy, S.N. (2017). The metabolism and significance of homocysteine in nutrition and health. *Nutrition & Metabolism*, 14, 78.

Laurent, L., Wong, E., Li, G., Huynh, T., Tsirigos, A., Ong, C.T., Low, H.M., Kin Sung, K.W., Rigoutsos, I., Loring, J., Wei, C.-L. (2010). Dynamic changes in the human methylome during differentiation. *Genome Research*, 20(3), 320–331.

Lazar, V., Ditu, L.-M., Pircalabioru, G.G., Picu, A., Petcu, L., Cucu, N., Chifiriuc, M.C. (2019). Gut microbiota, host organism, and diet trialogue in diabetes and obesity. *Frontiers in Nutrition*, 6, 21.

Li, D.Y., Brooke, B., Davis, E.C., Mecham, R.P., Sorensen, L.K., Boak, B.B., Eichwald, E., Keating, M.T. (1998). Elastin is an essential determinant of arterial morphogenesis. *Nature*, 393(6682), 276–280.

Linke, W.A. (2018). Titin gene and protein functions in passive and active muscle. *Annual Review of Physiology*, 80, 389–411.

Liu, X., Zhao, Y., Pawlyk, B., Damaser, M., Li, T. (2006). Failure of elastic fiber homeostasis leads to pelvic floor disorders. *The American Journal of Pathology*, 168(2), 519–528.

Longo, V. and Pelloso, Y. (2018). *Le régime de longévité*. Actes Sud, Arles.

Ludeman, L., Warren, B.F., Shepherd, N.A. (2002). The pathology of diverticular disease. *Best Practice & Research. Clinical Gastroenterology*, 16(4), 543–562.

Luisier, A.-C., Petitpierre, G., Ferdenzi, C., Clerc Bérod, A., Giboreau, A., Rouby, C., Bensafi, M. (2015). Odor perception in children with autism spectrum disorder and its relationship to food neophobia. *Frontiers in Psychology*, 6, 1830.

Maguire, D., Neytchev, O., Talwar, D., McMillan, D., Shiels, P.G. (2018). Telomere homeostasis: Interplay with magnesium. *International Journal of Molecular Sciences*, 19(1), E157.

Manisalidis, I., Stavropoulou, E., Stavropoulos, A., Bezirtzoglou, E. (2020). Environmental and health impacts of air pollution: A review. *Frontiers in Public Health*, 8, 14.

Maria, A.T.J., Bourgier, C., Martinaud, C., Borie, R., Rozier, P., Rivière, S., Crestani, B., Guilpain, P. (2020). De la fibrogenèse à la fibrose : mécanismes physiopathologiques et présentations cliniques. *La Revue de médecine interne*, 41(5), 325–329.

Maurice, J. and Manousou, P. (2018). Non-alcoholic fatty liver disease. *Clinical Medicine*, 18(3), 245–250.

Mesnage, R. and Antoniou, M.N. (2017). Facts and fallacies in the debate on glyphosate toxicity. *Frontiers in Public Health*, 5, 316.

Miall, R.C., Rosenthal, O., Ørstavik, K., Cole, J.D., Sarlegna, F.R. (2019). Loss of haptic feedback impairs control of hand posture: A study in chronically deafferented individuals when grasping and lifting objects. *Experimental Brain Research*, 237(9), 2167–2184.

Mitidieri, E., Cirino, G., d'Emmanuele di Villa Bianca, R., Sorrentino, R. (2020). Pharmacology and perspectives in erectile dysfunction in man. *Pharmacology & Therapeutics*, 208, 107493.

Mlodinow, L. (2018). *Elastic: Flexible Thinking in a Time of Change.* Pantheon, Paris.

Moore, J. and Thibeault, S. (2012). Insights into the role of elastin in vocal fold health and disease. *Journal of Voice: Official Journal of the Voice Foundation*, 26(3), 269–275.

Moreno Fernández-Ayala, D.J., Navas, P., López-Lluch, G. (2020). Age-related mitochondrial dysfunction as a key factor in COVID-19 disease. *Experimental Gerontology*, 142, 111147.

Moulin, L., Cenizo, V., Antu, A.N., André, V., Pain, S., Sommer, P., Debret, R. (2017). Methylation of LOXL1 promoter by DNMT3A in aged human skin fibroblasts. *Rejuvenation Research*, 20(2), 103–110.

Munn, L.L. (2015). Mechanobiology of lymphatic contractions. *Seminars in Cell & Developmental Biology*, 38, 67–74.

Okla, M., Kim, J., Koehler, K., Chung, S. (2017). Dietary factors promoting brown and beige fat development and thermogenesis. *Advances in Nutrition (Bethesda, MD)*, 8(3), 473–483.

Penny, D. (2015). Epigenetics, Darwin, and Lamarck. *Genome Biology and Evolution*, 7(6), 1758–1760.

Peters, C.H., Sharpe, E.J., Proenza, C. (2020). Cardiac pacemaker activity and aging. *Annual Review of Physiology*, 82(1), 21–43.

Pierron, D., Pereda-Loth, V., Mantel, M., Moranges, M., Bignon, E., Alva, O., Kabous, J., Heiske, M., Pacalon, J, David, R. et al. (2020). Smell and taste changes are early indicators of the COVID-19 pandemic and political decision effectiveness. *Nature Communications*, 11(1), 5152.

Poillot, P., O'Donnell, J., O'Connor, D.T., Ul Haq, E., Silien, C., Tofail, S.A.M., Huyghe, J.M. (2020). Piezoelectricity in the intervertebral disc. *Journal of Biomechanics*, 102, 109622.

Povedano, J.M., Martinez, P., Serrano, R., Tejera, Á., Gómez-López, G., Bobadilla, M., Flores, J.M., Bosch, F., Blasco, M.A. (2018). Therapeutic effects of telomerase in mice with pulmonary fibrosis induced by damage to the lungs and short telomeres. *eLife*, 7, e31299.

Qa'aty, N., Vincent, M., Wang, Y., Wang, A., Mitts, T.F., Hinek, A. (2015). Synthetic ligands of the elastin receptor induce elastogenesis in human dermal fibroblasts via activation of their IGF-1 receptors. *Journal of Dermatological Science*, 80(3), 175–185.

Reynaud, C., Ferreras, L., Di Mauro, P., Kan, C., Croset, M., Bonnelye, E., Pez, F., Thomas, C., Aimond, G., Karnoub, A.E., Brevet, M., Clézardin, P. (2017). Lysyl oxidase is a strong determinant of tumor cell colonization in bone. *Cancer Research*, 77(2), 268–278.

Robert, L. and Labat-Robert, J. (2015). Longevity and aging: Role of genes and of the extracellular matrix. *Biogerontology*, 16(1), 125–129.

van Rooij, E., Sutherland, L.B., Thatcher, J.E., DiMaio, J.M., Naseem, R.H., Marshall, W.S., Hill, J.A., Olson, E.N. (2008). Dysregulation of microRNAs after myocardial infarction reveals a role of miR-29 in cardiac fibrosis. *Proceedings of the National Academy of Sciences of the United States of America*, 105(35), 13027–13032.

Rowan, S., Bejarano, E., Taylor, A. (2018). Mechanistic targeting of advanced glycation end-products in age-related diseases. *Biochimica et Biophysica Acta. Molecular Basis of Disease*, 1864(12), 3631–3643.

Roy, M., Grattard, V., Dinet, C., Soares, A.V., Decavel, P., Sagawa, Y.J. (2020). Nordic walking influence on biomechanical parameters: A systematic review. *European Journal of Physical and Rehabilitation Medicine*, 56(5), 607–615.

Ruegsegger, G.N. and Booth, F.W. (2018). Health benefits of exercise. *Cold Spring Harbor Perspectives in Medicine*, 8(7), a029694.

Sá, O., Lopes, N., Alves, M., Caran, E. (2018). Effects of glycine on collagen, PDGF, and EGF expression in model of oral mucositis. *Nutrients*, 10(10), 1485.

Scheiber, I., Dringen, R., Mercer, J.F.B. (2013). Copper: Effects of deficiency and overload. *Metal Ions in Life Sciences*, 13, 359–387.

Schlötzer-Schrehardt, U., Pasutto, F., Sommer, P., Hornstra, I., Kruse, F.E., Naumann, G.O.H., Reis, A., Zenkel, M. (2008). Genotype-correlated expression of lysyl oxidase-like 1 in ocular tissues of patients with pseudoexfoliation syndrome/glaucoma and normal patients. *The American Journal of Pathology*, 173(6), 1724–1735.

Scialo, F., Daniele, A., Amato, F., Pastore, L., Matera, M.G., Cazzola, M., Castaldo, G., Bianco, A. (2020). ACE2: The major cell entry receptor for SARS-CoV-2. *Lung*, 198(6), 867–877.

Servan-Schreiber, D. and Dessert, S. (2011). *Anticancer : les gestes quotidiens pour la santé du corps et de l'esprit*, New edition (revised and extended). Pocket, Paris.

Servigne, P. and Chapelle, G. (2019). *L'entraide : l'autre loi de la jungle*. Éditions les Liens qui libèrent, Paris.

Shahar, B. (2020). New developments in emotion-focused therapy for social anxiety disorder. *Journal of Clinical Medicine*, 9(9), 2918.

Shapiro, S.D., Endicott, S.K., Province, M.A., Pierce, J.A., Campbell, E.J. (1991). Marked longevity of human lung parenchymal elastic fibers deduced from prevalence of D-aspartate and nuclear weapons-related radiocarbon. *Journal of Clinical Investigation*, 87(5), 1828–1834.

Simmonds, M.S.J., Howes, M.-J., Irving, J. (2018). *Plantes médicinales essentielles : des pharmacopées occidentale, chinoise et indienne*. Ulmer, Paris.

Skiada, A., Lass-Floerl, C., Klimko, N., Ibrahim, A., Roilides, E., Petrikkos, G. (2018). Challenges in the diagnosis and treatment of mucormycosis. *Medical Mycology*, 56(suppl_1), 93–101.

Sommer, P., Gleyzal, C., Guerret, S., Etienne, J., Grimaud, J.A. (1992). Induction of a putative laminin-binding protein of *Streptococcus gordonii* in human infective endocarditis. *Infection and Immunity*, 60(2), 360–365.

Spitz, J. (1938). L'Homme élastique [Online]. Available at: https://fr.wikipedia.org/w/index.php?title=L%27Homme_%C3%A9lastique&oldid=193478385 [Accessed 23 May 2022].

Sprott, H., Müller, A., Heine, H. (1997). Collagen crosslinks in fibromyalgia. *Arthritis & Rheumatism*, 40(8), 1450–1454.

Sun, H., Calabrese, E.J., Lin, Z., Lian, B., Zhang, X. (2020). Similarities between the Yin/Yang doctrine and hormesis in toxicology and pharmacology. *Trends in Pharmacological Sciences*, 41(8), 544–556.

Tansey, E.A. and Johnson, C.D. (2015). Recent advances in thermoregulation. *Advances in Physiology Education*, 39(3), 139–148.

Tenze, E. (2021). *E per cieli le scienze: la scienza di Dante*. Alighieri. Clown Bianco Edizioni, Ravenne.

Thieulin, C., Pailler-Mattei, C., Abdouni, A., Djaghloul, M., Zahouani, H. (2020). Mechanical and topographical anisotropy for human skin: Ageing effect. *Journal of the Mechanical Behavior of Biomedical Materials*, 103, 103551.

Torbet, J., Malbouyres, M., Builles, N., Justin, V., Roulet, M., Damour, O., Oldberg, A., Ruggiero, F., Hulmes, D.J.S. (2007). Tissue engineering of the cornea: Orthogonal scaffold of magnetically aligned collagen lamellae for corneal stroma reconstruction. *Annual International Conference of the IEEE Engineering in Medicine and Biology Society. IEEE Engineering in Medicine and Biology Society. Annual International Conference*, 2007, 6400.

Tort, P. (2015). L'effet Darwin. Sélection naturelle et naissance de la civilisation. *artescience*, 26 November [Online]. Available at: https://artescience.wordpress.com/2015/11/26/leffet-darwin-selection-naturelle-et-naissance-de-la-civilisation/.

Tort, P., Joli, M., Ollier, F., Paradis, C. (2020). *Capitalisme ou civilisation : entretiens avec Michel Joli, Fabien Ollier et Clément Paradis*. Éditions Gruppen, Paris.

Touillet, A., Peultier-Celli, L., Nicol, C., Jarrassé, N., Loiret, I., Martinet, N., Paysant, J., De Graaf, J.B. (2018). Characteristics of phantom upper limb mobility encourage phantom-mobility-based prosthesis control. *Scientific Reports*, 8(1), 15459.

Vallet, S.D. and Ricard-Blum, S. (2019). Lysyl oxidases: From enzyme activity to extracellular matrix cross-links. *Essays in Biochemistry*, 63(3), 349–364.

Verschuere, S., Navassiolava, N., Martin, L., Nevalainen, P.I., Coucke, P.J., Vanakker, O.M. (2021). Reassessment of causality of ABCC6 missense variants associated with pseudoxanthoma elasticum based on Sherloc. *Genetics in Medicine: Official Journal of the American College of Medical Genetics*, 23(1), 131–139.

Wahart, A., Hocine, T., Albrecht, C., Henry, A., Sarazin, T., Martiny, L., El Btaouri, H., Maurice, P., Bennasroune, A., Romier-Crouzet, B., Blaise, S., Duca, L. (2019). Role of elastin peptides and elastin receptor complex in metabolic and cardiovascular diseases. *The FEBS Journal*, 286(15), 2980–2993.

Weidner, C.I., Lin, Q., Koch, C.M., Eisele, L., Beier, F., Ziegler, P., Bauerschlag, D.O., Jöckel, K.-H., Erbel, R., Mühleisen, T.W. et al. (2014). Aging of blood can be tracked by DNA methylation changes at just three CpG sites. *Genome Biology*, 15(2), R24.

Yamauchi, M., Barker, T.H., Gibbons, D.L., Kurie, J.M. (2018). The fibrotic tumor stroma. *The Journal of Clinical Investigation*, 128(1), 16–25.

Younis, J.S., Abassi, Z., Skorecki, K. (2020). Is there an impact of the COVID-19 pandemic on male fertility? The ACE2 connection. *American Journal of Physiology-Endocrinology and Metabolism*, 318(6), E878–E880.

Zhang, R., Lao, L., Ren, K., Berman, B.M. (2014). Mechanisms of acupuncture-electroacupuncture on persistent pain. *Anesthesiology*, 120(2), 482–503.

Zhang, W.-B., Wang, G.-J., Fuxe, K. (2015). Classic and modern meridian studies: A review of low hydraulic resistance channels along meridians and their relevance for therapeutic effects in traditional Chinese medicine. *Evidence-Based Complementary and Alternative Medicine: eCAM*, 2015, 410979.

Pascal Sommer's bibliography

Selection of publications in the field with a reading committee

Annovi, G., Boraldi, F., Moscarelli, P., Guerra, D., Tiozzo, R., Parma, B., Sommer, P., Quaglino, D. (2012). Heparan sulfate affects elastin deposition in fibroblasts cultured from donors of different ages. *Rejuvenation Research*, 15(1), 22–31.

Behmoaras, J., Slove, S., Seve, S., Vranckx, R., Sommer, P., Jacob, M.-P. (2008). Differential expression of lysyl oxidases LOXL1 and LOX during growth and aging suggests specific roles in elastin and collagen fiber remodeling in rat aorta. *Rejuvenation Research*, 11(5), 883–889.

Boizot, J., Minville-Walz, M., Reinhardt, D.P., Bouschbacher, M., Sommer, P., Sigaudo-Roussel, D., Debret, R. (2022). FBN2 silencing recapitulates hypoxic conditions and induces elastic fiber impairment in human dermal fibroblasts. *International Journal of Molecular Sciences*, 23(3), 1824.

Boraldi, F., Annovi, G., Tiozzo, R., Sommer, P., Quaglino, D. (2010). Comparison of ex vivo and in vitro human fibroblast ageing models. *Mechanisms of Ageing and Development*, 131(10), 625–635.

Borel, A., Eichenberger, D., Farjanel, J., Kessler, E., Gleyzal, C., Hulmes, D.J., Sommer, P., Font, B. (2001). Lysyl oxidase-like protein from bovine aorta. Isolation and maturation to an active form by bone morphogenetic protein-1. *The Journal of Biological Chemistry*, 276(52), 48944–48949.

Bouez, C., Reynaud, C., Noblesse, E., Thépot, A., Gleyzal, C., Kanitakis, J., Perrier, E., Damour, O., Sommer, P. (2006). The lysyl oxidase LOX is absent in basal and squamous cell carcinomas and its knockdown induces an invading phenotype in a skin equivalent model. *Clin. Cancer Res.*, 12(5), 1463–1469.

Bouras, M., Tabone, E., Sommer, P., Bertholon, J., Droz, J.P., Benahmed, M. (2000). Detection of Smad4 gene mutation in Seminoma germ-cell tumors. *Cancer Research*, 15, 922–928.

Brune, T., Borel, A., Gilbert, T.W., Franceschi, J.P., Badylak, S.F., Sommer, P. (2007). In vitro comparison of human fibroblasts from intact and ruptured ACL for use in tissue engineering. *Eur. Cell. Mater.*, 14, 78–90.

Cenizo, V., André, V., Reymermier, C., Sommer, P., Damour, O., Perrier, E. (2006). LOXL as a target to increase the elastin content in adult skin: A dill extract induces the LOXL gene expression. *Experimental Dermatology*, 15(8), 574–581.

Chifflot, S., Sommer, P., Stussi-Garaud, C., Hirth, L. (1979). Identification of a soluble replicase in healthy and alfalfa mosaic virus infected tobacco plants. *Annales de phytopathologie*, 11, 568–569.

Chifflot, S., Sommer, P., Stussi-Garaud, C., Hirth, L. (1980). Replication of alfalfa mosaic virus RNA: Evidence for a soluble replicase in healthy and virus infected tobacco leaves. *Virology*, 100, 91–100.

Claus, S., Fischer, J., Mégarbané, H., Mégarbané, A., Jobard, F., Debret, R., Peyrol, S., Saker, S., Devillers, M., Sommer, P., Damour, O. (2008). A p.C217R mutation in fibulin-5 from cutis laxa patients is associated with incomplete extracellular matrix formation in a skin equivalent model. *The Journal of Investigative Dermatology*, 128(6), 1442–1450.

Croce, M.A., Sammarco, R., Paolinelli, D.C., Boraldi, F., Gheduzzi, D., Damour, O., Sommer, P., Pasquali R.I., Tiozzo, R., Quaglino, D. (2004). The skin equivalent model: Developments and perspectives. *Microscopie*, 2, 31–36.

Cruz-Jentoft, A.J., Franco, A., Sommer, P., Baeyens, J.P., Jankowska, E., Maggi, A., Ponikowski, P., Rys, A., Szczerbinska, K., Michel, J.-P., Milewicz, A. (2009). Silver paper: The future of health promotion and preventive actions, basic research, and clinical aspects of age-related disease – A report of the European Summit on Age-Related Disease. *Aging Clinical and Experimental Research*, 21(6), 376–385.

Debret, R., Cenizo, V., Aimond, G., André, V., Devillers, M., Rouvet, I., Mégarbané, A., Damour, O., Sommer, P. (2010). Epigenetic silencing of lysyl oxidase-like-1 through DNA hypermethylation in an autosomal recessive cutis laxa case. *The Journal of Investigative Dermatology*, 130(11), 2594–2601.

Decitre, M., Gleyzal, C., Raccurt, M., Peyrol, S., Aubert-Foucher, E., Csiszar, K., Sommer, P. (1998). Lysyl oxidase-like protein localizes to sites of de novo fibrinogenesis in fibrosis and in the early stromal reaction of ductal breast carcinomas. *Laboratory Investigation: A Journal of Technical Methods and Pathology*, 78(2), 143–151.

Decorps, J., Saumet, J.L., Sommer, P., Sigaudo-Roussel, D., Fromy, B. (2014a). Effect of ageing on tactile transduction processes. *Ageing Research Reviews*, 13, 90–99.

Desmoulière, A., Darby, I., Monte-Alto Costa, A., Raccurt, M., Tuchweber, B., Sommer, P., Gabbiani, G. (1997). Extracellular matrix deposition, lysyl oxidase expression and myofibroblast differentiation during the initial stages of cholestatic fibrosis in the rat. *Laboratory Investigation*, 76, 765–778.

Dussart, C., Ngo, M.-A., Siranyan, V., Sommer, P. (2019a). *De la démocratie sanitaire à la démocratie en santé*. LEH édition, Bordeaux.

Dussart, C., Ngo, M.-A., Siranyan, V., Sommer, P. (2019b). *De la démocratie sanitaire à la démocratie en santé : acte du colloque qui s'est tenu à la MSHS de Nice, le 3 décembre 2018*. LEH édition, Bordeaux.

Farjanel, J., Sève, S., Borel, A., Sommer, P., Hulmes, D.J.S. (2005). Inhibition of lysyl oxidase activity can delay phenotypic modulation of chondrocytes in two-dimensional culture. *Osteoarthritis and Cartilage*, 13(2), 120–128.

Fhayli, W., Boëté, Q., Kihal, N., Cenizo, V., Sommer, P., Boyle, W.A., Jacob, M.-P., Faury, G. (2020a). Dill extract induces elastic fiber neosynthesis and functional improvement in the ascending aorta of aged mice with reversal of age-dependent cardiac hypertrophy and involvement of lysyl oxidase-like-1. *Biomolecules*, 10(2), E173.

Giampuzzi, M., Botti, G., Cilli, M., Gusmano, R., Borel, A., Sommer, P., Di Donato, A. (2001). Down-regulation of lysyl oxidase-induced tumorigenic transformation in NRK-49F cells characterized by constitutive activation of ras proto-oncogene. *The Journal of Biological Chemistry*, 276(31), 29226–29232.

Gindre, D., Peyrol, S., Raccurt, M., Sommer, P., Loire, R., Grimaud, J.A., Cordier, J.F. (1995). Fibrosing vasculitis in Wegener's granulomatosis: Ultrastructural and immunohistochemical analysis of the vascular lesions. *Virchows Archiv: An International Journal of Pathology*, 427(4), 385–393.

Jourdan-Le Saux, C., Gleyzal, C., Garnier, J.M., Peraldi, M., Sommer, P., Grimaud, J.A. (1994). Lysyl oxidase cDNA of myofibroblast from mouse fibrotic liver. *Biochemical and Biophysical Research Communications*, 199(2), 587–592.

Jourdan-Le Saux, C., Gleyzal, C., Raccurt, M., Sommer, P. (1997). Functional analysis of the lysyl oxidase promoter in myofibroblast-like clones of 3T6 fibroblast. *Journal of Cellular Biochemistry*, 64(2), 328–341.

Jourdan-Le Saux, C., Le Saux, O., Gleyzal, C., Sommer, P., Csiszar, K. (2000). The mouse lysyl oxidase-like 2 gene (mLOXL2) maps to chromosome 14 and is highly expressed in skin, lung and thymus. *Matrix Biology: Journal of the International Society for Matrix Biology*, 19(2), 179–183.

Kirschmann, D.A., Seftor, E.A., Fong, S.F.T., Nieva, D.R.C., Sullivan, C.M., Edwards, E.M., Sommer, P., Csiszar, K., Hendrix, M.J.C. (2002). A molecular role for lysyl oxidase in breast cancer invasion. *Cancer Research*, 62(15), 4478–4483.

Lehmann, N., Debret, R., Roméas, A., Magloire, H., Degrange, M., Bleicher, F., Sommer, P., Seux, D. (2009). Self-etching increases matrix metalloproteinase expression in the dentin-pulp complex. *J. Dent. Res.*, 88(1), 77–82.

Lorion, C., Faye, C., Maret, B., Trimaille, T., Régnier, T., Sommer, P., Debret, R. (2014). Biosynthetic support based on dendritic poly(L-lysine) improves human skin fibroblasts attachment. *Journal of Biomaterials Science*, Polymer Edition, 25(2), 136–149.

Mainzer, C., Remoué, N., Molinari, J., Rousselle, P., Barricchello, C., Lago, J.C., Sommer, P., Sigaudo-Roussel, D., Debret, R. (2018). In vitro epidermis model mimicking IGF-1-specific age-related decline. *Experimental Dermatology*, 27(5), 537–543.

Moulin, L., Cenizo, V., Antu, A.N., André, V., Pain, S., Sommer, P., Debret, R. (2017). Methylation of LOXL1 promoter by DNMT3A in aged human skin fibroblasts. *Rejuvenation Research*, 20(2), 103–110.

Musso, O., Sommer, P., Drouet, E., Cotte, L., Neyra, M., Chevallier, M., Grimaud, J.A. (1996). In situ detection of human cytomegalovirus DNA in gastrointestinal biopsies from AIDS patients by means of various PCR-derived methods. *Journal of Virological Methods*, 56, 125–137.

Noblesse, E., Cenizo, V., Bouez, C., Borel, A., Gleyzal, C., Peyrol, S., Jacob, M.-P., Sommer, P., Damour, O. (2004). Lysyl oxidase-like and lysyl oxidase are present in the dermis and epidermis of a skin equivalent and in human skin and are associated to elastic fibers. *The Journal of Investigative Dermatology*, 122(3), 621–630.

Ogier, J., Klein, J.P., Sommer, P., Frank, R.M. (1983). Identification and preliminary characterization of saliva interacting surface antigens of *Streptococcus mutans* by immunoblotting, ligand blotting and immunoprecipitation. *Infection and Immunity*, 45, 107–112.

Ogier, M., Pini, A.L., Sommer, P., Klein, J.P. (1989). Purification and characterization of the expression product of the sr gene of *Streptococcus mutans*. *Microbial Pathogenesis*, 6, 175–182.

Ogier, M., Schöller, M., Lepoivre, Y., Sommer, P., Pini, A.L., Klein, J.P. (1990). Complete nucleotide sequence of the SR gene of *Streptococcus mutans* OMZ 175. *FEMS Microbiological Letters*, 68, 223–228.

Palamakumbura, A.H., Sommer, P., Trackman, P.C. (2003). Autocrine growth factor regulation of lysyl oxidase expression in transformed fibroblasts. *The Journal of Biological Chemistry*, 278(33), 30781–30787.

Palamakumbura, A.H., Jeay, S., Guo, Y., Pischon, N., Sommer, P., Sonenshein, G.E., Trackman, P.C. (2004). The propeptide domain of lysyl oxidase induces phenotypic reversion of ras-transformed cells. *The Journal of Biological Chemistry*, 279(39), 40593–40600.

Pascual, G., Mendieta, C., Mecham, R.P., Sommer, P., Bellón, J.M., Buján, J. (2008). Down-regulation of lysyl oxydase-like in aging and venous insufficiency. *Histology and Histopathology*, 23(2), 179–186.

Pascual, G., Rodríguez, M., Mecham, R.P., Sommer, P., Buján, J., Bellón, J.M. (2009). Lysyl oxidase like-1 dysregulation and its contribution to direct inguinal hernia. *European Journal of Clinical Investigation*, 39(4), 328–337.

Peyrol, S., Raccurt, M., Gerard, F., Gleyzal, C., Grimaud, J.A., Sommer, P. (1997). Lysyl oxidase gene expression in the stromal reaction to in situ and invasive ductal breast carcinoma. *The American Journal of Pathology*, 150(2), 497–507.

Peyrol, S., Galateau-Salle, F., Raccurt, M., Gleyzal, C., Sommer, P. (2000). Selective expression of lysyl oxidase (LOX) in the stromal reactions of bronchopulmonary carcinomas. *Histology and Histopathology*, 15(4), 1127–1135.

Pez, F., Dayan, F., Durivault, J., Kaniewski, B., Aimond, G., Le Provost, G.S., Deux, B., Clézardin, P., Sommer, P., Pouysségur, J., Reynaud, C. (2011). The HIF-1-inducible lysyl oxidase activates HIF-1 via the Akt pathway in a positive regulation loop and synergizes with HIF-1 in promoting tumor cell growth. *Cancer Research*, 71(5), 1647–1657.

Reynaud, C., Gleyzal, C., Jourdan-Le Saux, C., Sommer, P. (1999). Comparative functional study of the lysyl oxidase promoter in fibroblasts, Ras-transformed fibroblasts, myofibroblasts and smooth muscle cells. *Cellular and Molecular Biology (Noisy-Le-Grand, France)*, 45(8), 1237–1247.

Reynaud, C., Baas, D., Gleyzal, C., Le Guellec, D., Sommer, P. (2008). Morpholino knockdown of lysyl oxidase impairs zebrafish development, and reflects some aspects of copper metabolism disorders. *Matrix Biology: Journal of the International Society for Matrix Biology*, 27(6), 547–560.

Schlötzer-Schrehardt, U., Pasutto, F., Sommer, P., Hornstra, I., Kruse, F.E., Naumann, G.O.H., Reis, A., Zenkel, M. (2008). Genotype-correlated expression of lysyl oxidase-like 1 in ocular tissues of patients with pseudoexfoliation syndrome/glaucoma and normal patients. *The American Journal of Pathology*, 173(6), 1724–1735.

Schöller, M., Klein, J.P., Sommer, P., Frank, R.M. (1982). Common antigens of streptococcal and non streptococcal oral bacteria: Isolation and biochemical characterization of the extracellular protein antigen. *Journal of General Microbiology*, 128, 2113–2129.

Schöller, M., Klein, J.P., Sommer, P., Frank, R.M. (1983a). Protoplast and cytoplasmic membrane preparation from *Streptococcus sanguis* and *Streptococcus mutans*. *Journal of General Microbiology*, 129, 3271–3279.

Schöller, M., Klein, J.P., Sommer, P., Frank, R.M. (1983b). Common antigens of streptococcal and non streptococcal oral bacteria: Characterization of wall-associated protein and comparison with extracellular protein antigens. *Infection and Immunity*, 40, 1186–1191.

Seve, S., Decitre, M., Gleyzal, C., Farjanel, J., Sergeant, A., Ricard-Blum, S., Sommer, P. (2002). Expression analysis of recombinant lysyl oxidase (LOX) in myofibroblast-like cells. *Connective Tissue Research*, 43(4), 613–619.

Sibon, I., Sommer, P., Lamaziere, J.M.D., Bonnet, J. (2005). Lysyl oxidase deficiency: A new cause of human arterial dissection. *Heart (British Cardiac Society)*, 91(5), e33.

Sommer, P., Gizy-Andriamantena, A., Stussi-Garaud, C. (1981). Heterogeneous chromatographic behaviour of soluble RNA replicase from healthy and virus infected tobacco leaves. *Journal of Virological Methods*, 3, 229–239.

Sommer, P., Klein, J.P., Frank, R.M. (1982). Lactate dehydrogenases from *Streptococcus mutans* OMZ 175. *Journal of Dental Research*, 62, 452.

Sommer, P., Klein, J.P., Frank, R.M. (1983a). Characterization of fructose 1-6 diphosphate dependent lactate dehydrogenase from *Streptococcus mutans* OMZ 175. *Journal of Dental Research*, 63, 534.

Sommer, P., Klein, J.P., Frank, R.M. (1983b). Purification of fructose 1-6 diphosphate dependent lactate dehydrogenase from *Streptococcus mutans* OMZ 175. *Journal of Dental Research*, 63, 533.

Sommer, P., Klein, J.P., Ogier, M., Frank, R.M. (1985a). Immunological study of lactate dehydrogenase from *Streptococcus mutans* and evidence of common antigenic domains with lactate dehydrogenases from lactic bacteria. *Infection and Immunity*, 51, 277–281.

Sommer, P., Klein, J.P., Schöller, M., Frank, R.M. (1985b). Lactate dehydrogenase from *Streptococcus mutans*. Purification, characterization and crossed antigenicity with lactate dehydrogenase from *Lactobacillus casei*, *Actinomyces viscosus,* and *Streptococcus sanguis*. *Infection and Immunity*, 47, 485–495.

Sommer, P., Bruyere, T., Ogier, M., Garnier, J.M., Jeltsch, J.M., Klein, J.P. (1987). Cloning of a salivary interacting component from *Streptococcus mutans*. *Journal of Bacteriology*, 169, 5167–5173.

Sommer, P., Gleyzal, C., Guerret, S., Etienne, J., Grimaud, J.A. (1992). Induction of a putative laminin-binding protein of *Streptococcus gordonii* in human infective endocarditis. *Infection and Immunity*, 60(2), 360–365.

Sommer, P., Gleyzal, C., Raccurt, M., Delbourg, M., Serrar, M., Joazeiro, P., Peyrol, S., Kagan, H., Trackman, P.C., Grimaud, J.A. (1993). Transient expression of lysyl oxidase by liver myofibroblasts in murine schistosomiasis. *Laboratory Investigation: A Journal of Technical Methods and Pathology*, 69(4), 460–470.

Streichenberger, N., Peyrol, S., Philit, F., Loire, R., Sommer, P., Cordier, J.F. (2001). Constrictive bronchiolitis obliterans. Characterisation of fibrogenesis and lysyl oxidase expression patterns. *Virchows Archiv: An International Journal of Pathology*, 439(1), 78–84.

Thomassin, L., Werneck, C.C., Broekelmann, T.J., Gleyzal, C., Hornstra, I.K., Mecham, R.P., Sommer, P. (2005). The pro-regions of lysyl oxidase and lysyl oxidase-like 1 are required for deposition onto elastic fibers. *The Journal of Biological Chemistry*, 280(52), 42848–42855.

Trivedy, C., Warnakulasuriya, K.A., Hazarey, V.K., Tavassoli, M., Sommer, P., Johnson, N.W. (1999). The upregulation of lysyl oxidase in oral submucous fibrosis and squamous cell carcinoma. *Journal of Oral Pathology & Medicine*, 28(6), 246–251.

Index

A

abdominal hernias, 32
ACE2, 45, 50, 61, 98
adenosine triphosphate (ATP), 37–39, 101, 105, 118
adrenal glands, 45, 80
advanced glycation end products, 47
aerobic
    bacteria, 51
    respiration, 37
aging, 4, 9, 10, 39, 48, 56, 65, 71, 79, 83–85, 88, 90, 91, 106–108, 143
alveolar mucus, 22
alveoli, 5, 6, 22, 31, 32, 34, 35, 98, 99
amino acids, 8, 91, 93, 94, 96
amputation, 26, 113, 139
anabolism, 30, 47, 67, 83
anaerobic
    bacteria, 51
    respiration, 37
ankylosing spondylitis, 67, 75
anosmia, 119
anti-reflux valves, 40, 41
antihypertensive, 43, 80

arthritis, 25, 85
asphyxiation, 35
asthma, 99
atheromatous plaque, 99
atherosclerosis, 25, 93, 99
augmented human, 72, 125
autistic disorders, 118
autoimmune response, 51
autonomous movements, 64, 114
autonomy, 3, 131, 133, 142, 143

B

balloon, 29, 48
basal membrane, 15
behavior, 9, 73, 125, 135, 136, 138, 142, 143
bilharzia, 7
biliary insufficiency, 90
biocompatible, 48
biofilms, 50
biomaterials (*see also* elastic), 20, 61, 72
    biomimetic, 72
biomimicry, 52
bioreactor, 50–52
bladder, 3, 30, 42, 48, 74, 114

blindness, 47, 59
blood (*see also* oxygenated blood)
   capillaries, 35
   circulation, 31, 36, 40, 45, 53, 107
   pressure, 22, 34, 42–45, 64, 80, 90, 98, 116
   vessels, 6, 24, 44, 47, 59, 61, 64, 65, 91, 113, 139
   volume, 42
body hybridization, 72
bone, 20, 23, 26, 27, 51, 67, 72, 73, 95, 110, 111, 141
   fracture, 26, 67
bone skin, 27
breast palpation, 24
brittle bone disease, 23
bronchial tubes, 31, 34, 98, 106
bronchiolitis, 5, 99
bronchitis, 5
brown fat, 117
burns, 9, 67, 73, 117

C

cadmium, 39
calcification, 23, 25, 26, 92, 108, 141
calcium, 25, 37, 46, 67, 72, 92, 93, 99
   stones, 25, 67
cancer, 5, 12, 24, 46, 79, 85, 91
Candida albicans, 50
candidiasis, 50
capability, 72
capital, 1, 4, 5, 15, 29, 67, 77, 93, 101, 107, 124, 125, 131, 132, 137, 143
carbon dioxide, 5, 30–32, 35, 62, 64, 97, 98
cardiac system, 32, 34
cardiovascular system, 42, 44, 86, 96
care pathway, 129
caregivers, 4, 123, 129, 133

cartilages, 57, 62, 112, 118
catabolism, 30, 37, 47, 67, 83, 116
catalyst, 25
celiac disease, 97
cellular respiration, 37
central nervous system, 64
cerebral spinal fluid, 74
cervix, 49, 105
chain of values, 129
chewing, 56, 119
childbirth, 48, 49, 105
chlordecone, 39
cholesterol, 22, 99
chromosomes, 77
chronic fatigue syndrome, 39, 101
cirrhosis, 46
clitoris, 43
cochlea, 60, 63, 118
collagen, 10, 12, 13, 15, 18–20, 23–26, 30, 32, 47–49, 57, 59, 60, 62, 65, 74, 91, 93–95, 98, 112, 119
collapse, 34, 101
colon, 48, 51
combustion, 30, 31
compliance, 32, 98, 125, 126, 135–137, 139–141, 143, 145
compression, 10, 64, 120, 132
connectin, 22
copper, 39, 92, 95, 101
cornea, 59
coronavirus, 6, 131
COVID-19, 5–7, 22, 23, 25, 39, 50, 52, 56, 61, 62, 71, 77, 81, 95, 127, 131
Creutzfeldt-Jakob, 46
cross-linking, 23, 25, 47, 48, 59, 80, 85, 87, 95, 97, 125, 126, 130, 140, 141, 145
culture, 12, 86, 97, 114, 139–141, 145

cutis laxa syndrome, 15–17,
  19, 32, 34, 41, 43, 48, 61, 75, 76,
  79, 81, 97, 127
  acquired, 75, 97, 174
cytokines, 6

D

dance, 57, 117, 118, 140
dark circles, 41
decline, 4, 104, 106, 107, 135
decrease, 20, 32, 35, 44, 49, 50, 65,
  68, 98, 101, 104, 107, 112
dermal mucus, 31
dermis, 10, 11, 13, 15, 49, 87, 88
diabetes, 43, 47, 53, 59, 79, 90, 91,
  96
diaphragm, 35, 49, 57, 117
digestion, 9, 40, 51, 86, 90
dill, 85–87, 89, 96
dioxygen, 30
disability, 4, 57, 131, 133, 141–143
diverticula, 41, 67
DNA, 12, 14, 37, 39, 40, 75, 77, 88
  medicinal, 75
dominant mutations, 17

E

echinococcosis, 7, 24
ecology, 138, 143
edemas, 41
Ehlers-Danlos, 18, 19, 23, 57, 132
  syndrome, 18, 23, 57, 132
elastic, 1, 3–6, 8, 10, 12–15, 18–21,
  23–25, 27, 29–32, 34, 36, 41, 42,
  45–50, 52, 53, 57, 59–65, 67–77,
  79, 80, 83–88, 90–93, 95–99, 101,
  103, 104, 106–112, 116–119,
  123–126, 129–144

biomimetic biomaterials, 73
capital, 4, 6, 8, 10, 15, 24, 34, 45,
  48, 50, 65, 67, 68, 70, 72, 73,
  75, 77, 79, 80, 83–86, 90, 91,
  93, 95, 96, 98, 99, 101, 104,
  106, 108, 109, 116, 117, 119,
  123–125, 129, 130, 135, 136,
  142, 143
fibers, 8, 10, 12, 14, 15, 18–21, 23,
  24, 27, 30–32, 42, 46–49, 53,
  57, 59, 60, 62, 65, 72, 74, 80,
  83–88, 91, 93, 95, 97, 98, 112,
  117, 126
system, 12, 20, 34, 36, 45, 67, 71,
  91, 92, 120, 131, 136, 137, 144
elasticity, 3–5, 9–11, 13, 15, 19, 20,
  22, 24, 29, 30, 32–35, 37, 41, 43,
  47, 50, 51, 53, 55, 56, 58–65, 68,
  69, 71, 72, 77, 83, 85, 86, 90, 93,
  97–100, 103, 104, 106, 109, 111,
  112, 115, 116, 118, 120, 121,
  123–127, 130, 132, 134, 135, 137,
  139–141, 144, 145
elastin, 14, 17, 19, 22–24, 34, 59, 79,
  80, 90, 91, 93–95, 97, 99, 141
embolism, 67, 99
embryo, 24, 75
emotion, 18, 83, 118
emphysema, 3, 5, 32, 67, 98, 99
empowerment, 129
enamel, 51
endocarditis, 5, 52
energy, 5, 30, 31, 35, 37, 39, 40, 45,
  47, 59, 62, 83, 99, 101, 105, 106,
  114, 118, 120, 138, 142, 145
environment, 24, 40, 50, 52, 63,
  77–79, 87, 99, 119, 125, 131, 135,
  137, 144
epidermis, 10–13, 15, 26

epididymis, 50
epigenetics, 78, 79, 87, 125, 140
   imprint, 79, 87
epiglottis, 57
erectile dysfunction, 43
erythropoietin, 43
ethics, 68, 135
evolution, 4, 12, 34, 49, 55, 72, 78, 80, 95, 104, 128–130, 135, 136, 140
   adaptive theory of, 78
   selective theory of, 78
expert patients, 127
expiration, 32, 98
extension, 10, 22, 23, 32, 49, 120, 124, 135
extinction, 125, 134, 137, 142

F

fascias, 48, 53, 57, 63, 65, 111–113
fatty acids, 37, 40, 41, 92, 99
fertility, 50
fibrocartilage, 65
fibromyalgia, 39, 65, 101
fibrosis, 5–7, 10, 20, 23–25, 32, 46, 48, 67, 71, 73, 80, 94, 98, 108, 135
fibulins, 14
first aid, 84
food, 9, 22, 25, 37, 39–41, 45, 51, 74, 77, 83, 92–95, 97, 99, 101, 119, 124, 136, 137, 142, 143
   transit, 40
force, 12, 22, 40, 41, 46, 77, 103, 120, 139
friendliness, 17, 138
friendly, 138, 139
frog, 31
fungicides, 39, 101

G

gametes, 50
genetics, 17, 78, 79, 87, 125

genome, 12, 17, 79
glaucoma, 59
glucose, 37, 105, 117
glutamine, 93, 94
gluten, 97, 136
glycation, 47, 59, 93, 96, 99
glycemia, 91, 106
glycine, 91, 93–96
glycogen, 46
glyphosate, 39, 95
gonarthrosis, 25
gravity, 16, 29, 30, 40, 90, 107, 110, 113
growth, 4, 15, 18, 19, 24, 26, 46, 51, 52, 65, 91, 93–95, 105–107, 111, 132, 135, 137, 141, 142
   hormone, 46, 91, 93, 105, 107
gums, 51, 62

H

haptic, 63, 104, 115
healing, 10, 11, 13, 78, 84, 90, 93
hearing, 3, 6, 58, 60, 61, 113, 115, 117, 118
heart, 3, 5, 12, 24, 30, 32, 34, 35, 40, 44, 51, 53, 64, 67, 85, 92, 98, 99, 104–108, 134
   failure, 3, 35
   rate, 44, 64, 105, 107, 108
   valves, 34, 40
heartbeats, 24
hemoglobin, 96
heparin, 52
hernias, 3
histones, 88
homocysteine, 96, 99
hyaluronic acid, 21, 62
hydrogenation, 99
hyoid bone, 111
hyperglycemia, 46
hyperhomocysteinemia, 96
hyperlaxed, 18, 57

hypertelia, 72
hypertelic, 140
hypertension, 44, 45, 80, 85
hypotension, 45, 80

I

IGF1, 46, 49, 91, 93, 107
immune system, 13, 52, 75
immunodepression, 50
infarction, 24, 99
inhalation, 32, 43
inhibitors of succinate
  dehydrogenase, 39
inner ear, 60
inputs, 40, 83, 136
insulin, 46, 49, 91
  like growth factor, 46, 91
intestinal
  villi, 41
  walls, 40, 41
intestines, 30, 44, 51, 52, 143
iron, 7, 92, 94, 106
ischemia, 99

J, K, L

jellyfish, 12, 23, 32
joints, 65, 72, 106, 109, 111
justice, 87, 139, 143
keloids, 13
kidneys, 42–45, 47, 53, 74, 80, 90, 97, 106
knowledge, 17, 19, 22, 64, 70, 80, 96, 108, 125–130, 132, 133, 138, 145
lactoferrin, 94
lactose, 96
Langer's lines, 10, 11
large intestine, 40, 41
Laron syndrome, 91, 141
larynx, 57, 118

law, 30, 32, 79, 86, 125, 126, 135, 137, 143, 145
  of gas diffusion, 30, 32
  of mutual aid, 125, 126
laxity, 49, 143
lens, 59, 60
leverage, 120, 121
ligaments, 20, 25, 49, 53, 56, 63, 71, 72, 84, 94, 108, 111–113, 119
lindane, 39
liver, 7, 20, 24, 26, 30, 45–47, 53, 71, 88, 89, 91, 94, 141
  failure, 7
longevity, 4, 8, 17, 77, 83, 90, 91, 95, 108, 111
LOX (lysyl oxidase), 25, 97
lungs, 3–6, 8, 10, 22, 30–32, 34, 35, 42, 44, 47, 52, 53, 61, 85, 107, 116, 117, 126, 128
lymph, 30, 40, 41, 64, 74, 98
lymphatic
  channels, 41, 64, 113
  system, 41, 106

M, N

macrophages, 13
magnesium, 39, 92
magnetic resonance imaging, 71
male infertility, 50
Marfan syndrome, 18, 76, 80
marjoram, 87, 89, 90
material, 9, 47, 51, 53, 79, 104, 110, 126, 127, 131, 135
mechanics, 8, 13, 20, 21, 23, 31, 37, 43, 50, 51, 53, 55, 56, 71, 91, 96, 106, 108, 116, 119, 125, 126, 131, 133, 135, 136
melanin, 12, 15
melanocytes, 12, 15
melanomas, 12

memory, 89, 93, 105
menisci, 72
Menkes, 76
menopause, 48, 86
mercury, 39
metabolism, 47, 50, 74, 83
microbiota, 50–52, 117
microfibrils, 12, 14, 23
mitochondrial diseases, 75
model, 18, 86, 90, 114, 139
morality, 126, 140–142
morbidity, 81, 97
motivation, 118
movements, 3, 10, 41, 51, 55–57, 60, 63, 65, 108, 109, 114, 115
mucormycosis, 52
mucus, 22, 31, 34, 98, 99
multiple sclerosis, 65, 75
mumps, 50
muscles, 22, 26, 34, 37, 39, 48, 49, 53, 57, 59, 60, 63, 65, 94, 95, 98, 105, 106, 109, 111–113, 116, 119, 120
  cramp, 37
  fibers, 22, 49
muscular dystrophy, 39
mussel, 12
mutual aid, 125, 133, 135, 137, 139–143, 145
myelin, 64, 65
myeloma, 75, 81
myopia, 60
myth of Prometheus, 26
nature, 11, 23, 34, 35, 37, 41, 44, 45, 60, 72, 74, 78, 89, 99, 108, 117, 124, 125, 133, 134, 137, 139–141, 143–145
necrosis, 52, 84
negative pressure, 30
nephrons, 42
nitric oxide, 62
non-alcoholic fatty liver disease, 46

O, P

obesity, 22, 35, 43, 97
obstetrics, 49
olfaction, 6, 56, 58, 61, 118
olfactory bulb, 61
opportunity cost, 142
orphan disease, 16, 19
osteoarthritis, 67, 85, 93
osteogenesis imperfecta syndrome, 23
osteoporosis, 25, 93
otherness, 123, 133
oxygen, 30–32, 35, 37, 39, 40, 43, 45, 50, 51, 62, 74, 90, 97, 98, 105–107
oxygenated blood, 32, 34
oxygenation, 32, 37, 67, 99, 105
oxytocin, 57, 77, 105
pain, 4–6, 25, 56, 63–65, 70, 105, 108, 111, 113, 116, 128, 142
parasympathetic, 44, 105, 107, 108
  system, 44, 105, 107, 108
pelvic organs, 48
penis, 43, 50, 141
perineal space, 47, 48
perineum, 49, 57, 117
perineurium, 64, 65
periosteum, 27
periostitis, 27
pesticides, 39, 43, 101
phantom limb, 113, 138
pharynx, 57, 112
phonation, 119
phosphate, 37, 95
phosphodiesterase inhibitors, 43
physical therapy, 107, 120
phytotherapy, 83, 84, 89
piezoelectric, 110
pituitary gland, 45, 80
plague, 5, 6
plastic surgery, 15
polymer, 136
population genetics, 135

pregnancy, 49, 91
pressure, 5, 10, 15, 22, 30, 32, 34, 40, 42, 44, 46, 47, 49–51, 53, 60, 62, 64, 98, 102, 107, 112, 114, 115, 117, 127, 129, 131, 138, 142, 143
procreation, 3
progeria syndrome, 18, 77
proline, 93, 94
proprioception, 62, 63, 89, 110, 112, 113, 115, 119
protein restriction, 90, 91
public health, 72, 129, 143
pulmonary
  alveoli, 31, 73, 99, 100, 135
  edema, 6, 35, 98
  fibrosis, 25, 32, 67, 98
  insufficiency, 35

Q, R

quality control system, 74
radiations, 20
radiotherapy, 5, 10
rare diseases, 17, 19, 128
recessive mutations, 17
reconstructed skin, 86
reconstructive surgery, 21, 73
red blood cells, 43
regeneration, 26
rehabilitation, 118, 120
relaxin, 49, 91
renal failure, 6
renin-angiotensin system, 45
repaired human, 72
resilience, 126, 131–135, 140, 141, 143, 145
respiration, 7, 22, 30, 34, 37, 40
retina, 47, 59, 75
reward, 105, 110, 118, 130, 139, 142
rheumatic arthritis, 25
rheumatoid arthritis, 3
ribose, 37

RNA
  drug therapy, 76
  vaccination, 77
  virus, 6, 50, 84
rubber, 11, 13, 53
  tree, 11

S, T

salamander, 26
salinity, 42, 50
SARS, 5–7, 17, 35, 44, 45, 50, 52, 61, 71, 76, 80, 95
scale, 20, 35, 50, 67, 70, 97, 117, 129, 133, 134, 139
scars, 5, 9, 10, 12, 13, 16
sedentarization, 124
sedentary, 41, 60, 81, 108, 137, 142
sense of smell, 61, 89
sensible eating, 83
sensoriality, 55, 56, 124
sexuality, 48, 50
silkworm, 11
singing, 44, 56, 57, 70, 107, 117
skin, 4, 8–13, 15, 16, 18–20, 31, 41, 43, 49, 52, 53, 56, 62, 63, 65, 71, 73–75, 80, 84–86, 88, 92, 94, 112–116
  flakes, 136
small intestine, 40–42, 51
sociability, 57, 119, 120
social networks, 126, 128, 130
socialization, 118
solar elastosis, 74, 80
sound wave, 56
Spanish flu, 5, 6
spermatozoa, 50
sphincters, 40
spider, 11
spinal cord, 64
splenomegaly, 71
staphylococci, 51
stem cell, 26, 27, 72, 73

stereotypes, 70, 123, 133
streptococci, 5, 51
stress, 6, 9, 11, 15, 44, 67, 69–71, 74, 75, 77–79, 81, 83, 93, 105, 108, 110, 111
stretch marks, 10, 49, 74
stroma of tumors, 24
sugars, 40, 45, 51, 92, 94, 96, 136
supravalvular aortic stenosis, 76
sweet pea, 24, 25, 80, 95
sympathetic system, 45, 57, 62, 105
symphysis pubis, 49
system, 7, 12, 17, 30, 41, 44, 45, 53, 62, 64, 65, 74, 75, 80, 98, 105, 112, 118, 120, 124, 126, 128, 130, 132, 135–137, 140, 145
   of well-being, 132
tactile, 15, 62
tapestry of the Apocalypse, 8
taste, 47, 62, 119, 138, 139
tattoo, 13
teeth, 25, 37, 56, 62, 92, 104, 119
telomeres, 77
tendons, 20, 63, 72, 84, 108, 111, 113
theory
   of aging, 39, 90
   of dietary restriction, 83
therapeutic wandering, 4, 17, 75, 114
thermodynamics, 145
tissue remodeling, 24
titin, 22, 23, 93
tongue, 57, 62, 112, 119
touch, 15, 56, 62, 63, 78, 115, 119
toxicity, 95, 96
toxins, 24, 25, 43
"transparent human", 71
tuberculosis, 5, 6
tympanum, 60

U, V

ultrasound, 71
ultraviolet rays, 74
umbelliferae, 87, 89
unhealthy diet, 83
urethra, 48, 50
urine, 42, 43, 45, 53, 56
uroscopy, 53
uterine muscles, 49
uterus, 48
vagal shock, 44
vagina, 48, 49
vaginal prolapse, 48
vagus nerve, 44, 57, 62, 107
valine, 93, 94
vas deferens, 50
vascular
   pressure, 34, 124
   rupture, 34
vasoconstriction, 44, 62, 120
vasodilation, 43, 44, 50, 62, 116, 120
ventricles, 40
vision, 3, 6, 8, 23, 53, 56, 58, 59, 63, 89, 109, 112, 113, 115, 119, 125, 132, 133, 144
visual aid, 113
vitamin, 92
vocal cords, 56, 57, 73, 99, 117, 118
voice, 17, 56, 57, 99, 118, 130
volume, 32, 42, 43, 73, 107

W, X, Y, Z

wall, 6, 31, 36, 41, 93, 97
war maiming, 70
waste, 30, 31, 40, 41, 47, 64, 74, 94, 106, 138
well-being, 67, 129, 132
Williams-Beuren syndrome, 18, 79, 141
wrinkles, 3, 4, 9, 10, 15, 73, 78
X and Y chromosomes, 40
yoga, 30, 44, 62, 107, 113, 114, 120
yogic breathing, 35
youth, 4, 15, 63, 90
zinc, 92

Other titles from

in

Health Engineering and Society

2023

BEYALA Laure
Digital Therapy: The New Age of Healthcare
(Technological Prospects and Social Applications Set – Volume 6)

2022

ASSAILLY Jean-Pascal
Child Psychology: Developments in Knowledge and Theoretical Models
(Health and Patients Set – Volume 3)

CHAABANE Sondès, COUSEIN Etienne, WIESER Philippe
Healthcare Systems: Challenges and Opportunities

PAGANELLI Céline, CLAVIER Viviane
Information Practices and Knowledge in Health
(Health and Patients Set – Volume 5)

SALGUES Bruno, BARNOUIN Jacques
The Covid-19 Crisis: From a Question of an Epidemic to a Societal Questioning
(Health and Patients Set – Volume 4)

2021

BÉRANGER Jérôme
Societal Responsibility of Artificial Intelligence: Towards an Ethical and Eco-responsible AI
(Technological Prospects and Social Applications Set – Volume 4)

CORDELIER Benoit, GALIBERT Olivier
Digital Health Communications
(Technological Prospects and Social Applications Set – Volume 5)

NADOT Michel
Discipline of Nursing: Three-time Knowledge

2020

BARBET Jacques, FOUCQUIER Adrien, THOMAS Yves
Therapeutic Progress in Oncology: Towards a Revolution in Cancer Therapy?
(Health and Patients Set – Volume 2)

MORILLON Laurent, CLAVIER Viviane
Health Research Practices in a Digital Context
(Health Information Set – Volume 4)

PAILLIART Isabelle
New Territories in Health
(Health Information Set – Volume 3)

2019

CLAVIER Viviane, DE OLIVEIRA Jean-Philippe
Food and Health: Actor Strategies in Information and Communication
(Health Information Set – Volume 2)

PIZON Frank
Health Education and Prevention
(Health and Patients Set – Volume 1)

2018

HUARD Pierre
The Management of Chronic Diseases: Organizational Innovation and Efficiency

PAGANELLI Céline
Confidence and Legitimacy in Health Information and Communication (Health Information Set – Volume 1)

2017

PICARD Robert
Co-design in Living Labs for Healthcare and Independent Living: Concepts, Methods and Tools

2015

BÉRANGER Jérôme
Medical Information Systems Ethics

Printed and bound by CPI Group (UK) Ltd, Croydon, CR0 4YY
01/10/2023

08123560-0001